International Conference on Image Optimization in Nuclear Medicine 2011

(OptiNM)

Journal of Physics: Conference Series Volume 317

Ayia Napa, Cyprus
23-26 March 2011

ISBN: 978-1-61839-288-6
ISSN: 1742-6588

Printed from e-media with permission by:

Curran Associates, Inc.
57 Morehouse Lane
Red Hook, NY 12571

Some format issues inherent in the e-media version may also appear in this print version.

Printed by Curran Associates, Inc. (2011)

For permission requests, please contact the Institute of Physics
at the address below.

Institute of Physics
Dirac House, Temple Back
Bristol BS1 6BE UK

Phone: 44 1 17 929 7481
Fax: 44 1 17 920 0979

techtracking@iop.org

Additional copies of this publication are available from:

Curran Associates, Inc.
57 Morehouse Lane
Red Hook, NY 12571 USA
Phone: 845-758-0400
Fax: 845-758-2634
Email: curran@proceedings.com
Web: www.proceedings.com

TABLE OF CONTENTS

IMAGE OPTIMISATION

IMAGE ANALYSIS

PATIENT DOSIMETRY

FUNCTIONAL IMAGING

SIMULATIONS

CLINICAL AND QUALITY MANAGEMENT

EDUCATION AND TRAINING OF MEDICAL PHYSICISTS IN NUCLEAR MEDICINE

PREFACE

OPTINM 2011

The International Conference on Image Optimisation in Nuclear Medicine was held at the Atlantica Aeneas Resort in Ayia Napa, Cyprus between 23–26 March 2011.

It was organised in the framework of the research project "Optimising Diagnostic Value in SPECT Myocardial Perfusion Imaging" (ΥΓΕΙΑ/ΔΥΓΕΙΑ/0308/11), funded by the Cyprus Research Promotion Foundation and the European Regional Development Fund, to present the highlights of the project, discuss the progress and results, and define future related goals.

The aim of this International Conference was to concentrate on image optimization approaches in Nuclear Medicine. Experts in the field of nuclear medicine presented their latest research results, exchanged experiences and set future goals for image optimisation while balancing patient dose and diagnostic value.

The conference was jointly organized by the Frederick Research Centre in Cyprus, the Department of Medical and Public Health Services of the Cyprus Ministry of Health, the Biomedical Research Foundation in Cyprus and the AGH University of Science and Technology in Poland. It was supported by the Cyprus Association of Medical Physics and Biomedical Engineering, and the Cyprus Society of Nuclear Medicine. The conference was held under the auspices of the European Federation of Organisations for Medical Physics and the European Association of Nuclear Medicine.

The conference scientific programme covered several important topics such as functional imaging; image optimization; quantification for diagnosis; justification; simulations; patient dosimetry, staff exposures and radiation risks; quality assurance and clinical audit; education, training and radiation protection culture; hybrid systems and image registration; and new and competing technologies. The programme consisted of 13 invited and keynote presentations as well as workshops, round table discussions and a number of scientific sessions. A total of 51 speakers presented their research and results to more than 150 participants from 14 countries. During the conference, exhibitors presented medical equipment used in nuclear medicine.

We gratefully acknowledge the financial support of the Cyprus Research Promotion Foundation, the European Regional Development Fund and the Cyprus Biomedical Research Foundation. Also, we appreciate the support of the various local sponsors listed in the conference programme.

We would like to express our sincere thanks and gratitude to the organising committee, the scientific committee and the supporting professional organizations for the success of the conference.

We also thank all of speakers for their excellent contributions, all the participants for their input, and the exhibitors for their valuable presentations.

Special thanks go to Demetris Kaolis, Maria Christofidou, Isabelle Chrysanthou, Charalambos Yiannakkaras, Ourania Demetriadou, and Elena Christofidou for their invaluable contribution to the conference.

The conference volume consists of 26 selected proceedings papers. We would like to thank all of the authors for their time and genuine efforts and the reviewers for their fruitful comments.

The Conference Chairpersons

Stelios Christofides and Yiannis Parpottas

Scientific Programme Committee

John Buscombe, United Kingdom
Carmel Caruana, Malta
Alberto del Guerra, Italy
Ourania Demetriadou, Cyprus
Jure Fettich, Slovenia
Chariklia Giannopoulou, Greece
Glenn Flux, United Kingdom
Savvas Frangos, Cyprus
Maria Lyra, Greece
Lesley Malone, Ireland
Paul Marsden, United Kingdom
Sören Mattsson, Sweden
Ajit Kumar Padhy, Singapore
Renato Padovani, Italy
Peter Sharp, United Kingdom
Istvan Szilvasi, Hungary
Marta Wasilewska-Radwanska, Poland

Invited Speakers

Demetris Apostolopoulos, Greece
John Buscombe, United Kingdom
Jure Fettich, Slovenia
Glenn Flux, United Kingdom
Christina Frankagi, Greece
Chariklia Giannopoulou, Greece
Bieke Lambert, Belgium
Michael Ljungberg, Sweden
Paul Marsden, United Kingdom
Nick Papathanasiou, Greece
Phivi Rondogianni, Greece
Anna Teresinska, Poland
Marta Wasilewska-Radwanska, Poland

Collaborating Organisations

Frederick Research Centre, Nicosia, Cyprus
Ministry of Health, Nicosia, Cyprus
Biomedical Research Foundation, Nicosia, Cyprus
AGH University of Science and Technology, Krakow, Poland

Supporting Organisations

Cyprus Association of Medical Physics (CAMPBE), Nicosia, Cyprus
Cyprus Society of Nuclear Medicine (CySNM), Limassol, Cyprus

Under the auspices of

European Association of Nuclear Medicine (EANM), Vienna, AU
European Federation of Organisations for Medical Physics (EFOMP), York, UK

Conference Chairpersons

Stelios Christofides and Yiannis Parpottas

Local Organising Committee

Isabelle Chrysanthou, Ourania Demetriadou, Demetris Kaolis, Prodromos Kaplanis, Charalambos Yiannakkaras

www.campbe.org/optinm

Tel.: +357 22 603136
Fax.: +357 22 603137
optinm@campbe.com

This Conference has received funding from the Cyprus Research Promotion Foundation and the European Regional Development Fund through the project ΥΓΕΙΑ/ΔΥΓΕΙΑ/0308/11: Optimising Diagnostic Value in SPECT Myocardial Perfusion Imaging.

The International Conference on Image Optimisation in Nuclear medicine is sponsored by:

Cyprus Telecommunications Authority
Address: P.O. Box 24929, Telecommunications Street, Nicosia, CY-1396, Cyprus
Phone: +357 22 70 10 00
Fax: +357 22 49 49 40
Web: http://www.cyta.com.cy/

Cyprus Broadcasting Corporation
Address: P.O. Box 24824, Nicosia, CY-1397, Cyprus
Phone: +357 22 862000
Fax: +357 22 317585
E-mail: rik@cybc.com.cy
Web: http://www.cybc.com.cy

Cyprus Tourism Organisation
Address: P.O. Box 24535, Nicosia, CY-1390, Cyprus
Phone: +357 22 691100
Fax: +357 22 338541
Web: http://www.visitcyprus.com/wps/portal

W.P. Wiseprojects ltd
Address: Ithakis 3, Apt. C2-001, Strovolos, Nicosia CY-2058, Cyprus
Phone: +357 22 519351
Fax: +357 22 493441
E-mail: demdepp@hotmail.com

Papaetis Medical Co Ltd
Address: Athalassa Avenue 75, Chapo Tower, Mazzanine Office 2, P.O. Box 25456, Nicosia, CY-2012, Cyprus
Phone: +357 22 204600
Fax: +357 22 314018
E-mail: tasoulla@papaetis.com

AGF Trading & Engineering Ltd
Address: Michalakis Kousoulides 26, Nicosia, CY-2024, Cyprus
Phone: +357 22 313900
Fax: +357 22 313144
Web: http://www.agfhealth.com

J.N. Christofides Trading Ltd
Address: Michalakopoulou 29A, P.O. Box 21093, Nicosia, CY-1075, Cyprus
Phone: +357 22 875257
Fax: + 357 22 764643
E-mail: kimonc@jnc.com.cy

Costas Christodoulou Ltd
Address: Theofani Theodotou 30, P.O.Box: 21032, Nicosia, CY-1500, Cyprus
Phone: +357 22 679843
Fax: +357 22 677087
E-mail: info@costaschristodoulou.com.cy
Web: www.costaschristodoulou.com.cy

MEDVISION The medical Solutions Company
Address: Gregori Afxentiou 5, Nicosia, CY-1096, Cyprus
Phone: +357 22 753591
Fax: +357 22 753597
E-mail: medvision@spidernet.com.cy

Scientronics Ltd Better Diagnostics Better Life
Address: Megalou Alaxandrou Avenue 7, Latsia, Nicosia, CY-2233 , Cyprus
Phone: +357 22 467880
Fax: +357 22 764614
E-mail: a.sarpetsas@scientronics.com

S. Stylianou Medisupplies
Address: Gonia Loizou Askani & Mopasa 2, Ekali, Limassol, CY-3110, Cyprus
Phone: +357 25 333530
Fax: +357 25 384540
E-mail: s-styl-medisup@cytanet.com.cy

This Conference has received funding from the Cyprus Research Promotion Foundation and the European Regional Development Fund through the project ΥΓΕΙΑ/ΔΥΓΕΙΑ/0308/11: Optimising Diagnostic Value in SPECT Myocardial Perfusion Imaging.

FREDERICK RESEARCH CENTRE

AGH — AGH University of Science and Technology Krakow Poland

IBE

International Conference

Image Optimisation in Nuclear Medicine

23rd to 26th of March 2011

Atlantica Aeneas Resort & Spa ***, Ayia Napa, Cyprus**

Tel.: +357 22 603136, Fax.: +357 22 603137, email: optinm@campbe.com, www.campbe.org/optinm

Final Programme

REGISTRATION ***Hotel, lobby***	
Wednesday 23rd of March 2011	13:30 - 18:00
Thursday 24th of March 2011	08:00 -18:00

WEDNESDAY, 23rd of March 2011		
15:00 – 16:30 **Athena**	**OPENING CEREMONY** ***Chairs:*** *Stelios Christofides (Cyprus), Yiannis Parpottas (Cyprus)* *Presenter: Yiannis Parpottas*	
	Welcome by the Conference Chairpersons	Stelios Christofides
	Welcome by the Representative of the Director General of the Research Promotion Foundation	Rebecca Chrysafi
	Welcome by the President of the Cyprus Association of Medical Physics and Biomedical Engineering	Charalambos Yiannakkaras
	Welcome by the President of the Cyprus Society of Nuclear Medicine	Ourania Demetriadou
	Welcome by the President of the Biomedical Research Foundation	Ourania Demetriadou
	Welcome by the Representative of the Council President of Frederick University	Stavros Christofi
	Welcome by the Representative of the AGH University of Science and Technology	Marta Wasilewska-Radwanska
	Welcome by the President of the European Federation of Organisations for Medical Physics	Stelios Christofides
	Welcome by the Secretary of the European Association of Nuclear Medicine	Savvas Frangos
	Welcome by the Director General of the Ministry of Health of Cyprus	Dionysios Mavronicolas
16:30 – 17:00	**Coffee Break**	

This Conference has received funding from the Cyprus Research Promotion Foundation and the European Regional Development Fund through the project ΥΓΕΙΑ/ΔΥΓΕΙΑ/0308/11: Optimising Diagnostic Value in SPECT Myocardial Perfusion Imaging.

ΕΥΡΩΠΑΪΚΗ ΕΝΩΣΗ

ΚΥΠΡΙΑΚΗ ΔΗΜΟΚΡΑΤΙΑ

ΔΙΑΡΘΡΩΤΙΚΑ ΤΑΜΕΙΑ

ΔΕΣΜΗ 2008

WEDNESDAY, 23rd of March 2011		
17:00 – 18:00 **Athena**	**CONFERENCE OPENING LECTURE** ***Chairs:*** *Stelios Christofides (Cyprus), Yiannis Parpottas (Cyprus)*	
	OL The State-of-the-Art of the PET-MRI Hybrid Systems	Paul Marsden, UK

19:00 – 21:00	**Conference Cocktail Reception**

THURSDAY 24th of March 2011		
08:00 – 10:00 **Athena**	SCIENTIFIC SESSION 1 ***Image Optimisation*** ***Chairs:*** *Savvas Frangos (Cyprus), Antonis Stefanoyiannis (Greece)*	
08:00 – 08:20	**SS1-1** Evaluation of two approaches to motion-corrected PET image reconstruction	Irene Polycarpou, UK
08:20 – 08:40	**SS1-2** Reconstruction of Linear Kinetic Parameters Directly From Projection PET Data	Georgios Angelis, UK
08:40 – 09:00	**SS1-3** Acquisition parameters for optimization of Bremsstrahlung (Yttrium - 90)	Monika Kempinska, Poland
09:00 – 09:20	**SS1-4** Justification of administered dose level in brain perfusion imaging with ^{99m}Tc-HMPAO	Antonis Stefanoyiannis, Greece
09:20 – 09:40	**SS1-5** Optimisation of SPECT reconstruction with OSEM and distance-dependent system resolution model (DRM) in relation to data acquisition parameters in clinical studies – Effects of count statistics and angular sampling	Lefteris Livieratos, UK
09:40 – 10:00	Discussion	
10:00 – 10:30	**Coffee Break**	

This Conference has received funding from the Cyprus Research Promotion Foundation and the European Regional Development Fund through the project ΥΓΕΙΑ/ΔΥΓΕΙΑ/0308/11: Optimising Diagnostic Value in SPECT Myocardial Perfusion Imaging.

ΕΥΡΩΠΑΪΚΗ ΕΝΩΣΗ ΚΥΠΡΙΑΚΗ ΔΗΜΟΚΡΑΤΙΑ ΔΙΑΡΘΡΩΤΙΚΑ ΤΑΜΕΙΑ ΔΕΣΜΗ 2008

FREDERICK RESEARCH CENTRE

AGH

IBE

THURSDAY 24th of March 2011			
10:30 – 13:10 **Athena**	**SCIENTIFIC SESSION 2** ***Image Analysis*** ***Chairs:*** *Jure Fettich (Slovenia), Ourania Demetriadou (Cyprus)*		
10:30 – 11:30	**IL-1**	The Role of SPECT/CT Hybrid Imaging on evaluation of tumors	Jure Fettich, Slovenia
11:30 – 11:50	**SS2-1**	A comparison between conventional manual ROI method and an automatic algorithm for semiquantitative analysis of SPECT studies	Laura Pagan, Italy
11:50 – 12:10	**SS2-2**	A combined [11C]-(+)-PHNO Positron Emission Tomography (PET) and Diffusion Tensor Imaging (DTI) study to measure the Amphetamine-induced Dopamine release within the functional subdivisions of the Basal Ganglia	Andri Tziortzi, UK
12:10 – 12:30	**SS2-3**	DaTSCAN studies: Differences in classification using 2D and 3D analysis	Jim O' Doherty, UK
12:30 – 12:50	**SS2-4**	Investigation of the Effects of Attenuation Correction on the Compatibility of two SPECT Systems with the use of Segmentation through Registration	Marios Lampaskis, Cyprus
12:50 – 13:10	**SS2-5**	Diagnosis Improvement Based on Image Processing and Dosimetric Aspects in Low Extremities Inflammation Scintigraphy	Maria Lyra, Greece
13:10 – 14:00	**Lunch**		
14:00 – 16:30 **Athena**	**SCIENTIFIC SESSION 3** ***Patient Dosimetry*** ***Chairs:*** *Sören Mattsson (Sweden), Maria Lyra (Greece)*		
14:00 – 14:30	**KN-1**	Current activities in the ICRP concerning estimation of radiation doses to patients from radiopharmaceuticals	Sören Mattsson, Sweden
14:30 – 14:50	**SS3-1**	Impact of low dose CT scan in dual timepoint investigations: A phantom study	Mille Micheelsen, Denmark
14:50 – 15:10	**SS3-2**	Safety of Radiosynovectomy with Yttrium-90	Monika Kempinska, Poland
15:10 – 15:30	**SS3-3**	Dosimetry of Exendin-4 Based Radiotracer for Glucagonlike Peptide-1 Receptor Imaging – An Initial Report	Monika Tomaszuk, Poland
15:30 – 15:50	**SS3-4**	Radioiodine Therapy of Differentiated Thyroid Cancer - The Protocol of Treatment and its Clinical Practicability	Daniela Neamtu, Romania
15:50 – 16:10	**SS3-5**	Effective doses to family member of patients treated with radioiodine-131	Marina Zdraveska Kocovska, FYROM
16:10 – 16:30	**SS3-6**	Scintigraphic imaging data in skeletal Metastasis: Treatment and Monte Carlo Simulation, Re-186 and Sm-153 dosimetry	Maria Andreou, Greece
16:30 – 17:00	**Coffee Break**		

This Conference has received funding from the Cyprus Research Promotion Foundation and the European Regional Development Fund through the project ΥΓΕΙΑ/ΔΥΓΕΙΑ/0308/11: Optimising Diagnostic Value in SPECT Myocardial Perfusion Imaging.

ΕΥΡΩΠΑΪΚΗ ΕΝΩΣΗ

ΚΥΠΡΙΑΚΗ ΔΗΜΟΚΡΑΤΙΑ

ΔΙΑΡΘΡΩΤΙΚΑ ΤΑΜΕΙΑ

ΔΕΣΜΗ 2008

THURSDAY 24th of March 2011		
17:00 – 19:00 **Athena**	**SPECIAL SESSION** ***Radionuclide Therapy*** ***Chairs:*** *John Buscombe (United Kingdom), Glenn Fkux (United Kingdom)*	
17:00 – 17:30	**SS-1** Clinical perspective of dosimetry in solid tumours treated with antibodies or peptides	John Buscombe, UK
17:30 – 18:00	**SS-2** Yttrium-90 microspheres for liver tumours: practical aspects	Bieke Lambert, Belgium
18:00 – 18:30	**SS-3** The Role of Radiation Dosimetry in Radionuclide Therapy	Glenn Flux, UK
18:30 – 19:00	Discussion	

FRIDAY 25th of March 2011		
08:00 – 10:00 **Athena**	**WORKSHOP 1** ***Optimising Diagnostic Value in SPECT Myocardial Perfusion Imaging*** ***Chairs:*** *Ourania Demetriadou (Cyprus), Marta Wasilewska-Radwanska (Poland)*	
08:00 – 09:00	**IL2** Gated SPECT imaging in acute myocardial infarction	Anna Teresinska, Poland
09:00 – 09:20	**W1-1** Optimising Diagnostic Value in SPECT Myocardial Perfusion Imaging	Yiannis Parpottas, Cyprus
09:20 - 09:40	**W1-2** Prototype of a dynamic heart phantom for optimizing diagnostic value in SPECT myocardial perfusion imaging	Marta Wasilewska-Radwanska, Poland
09:40 – 10:00	**W1-3** Optimised SPECT Myocardial Perfusion Images with a prototype dynamic heart phantom	Ourania Demetriadou, Cyprus
10:00 – 10:30	**Coffee Break**	
10:30 – 11:50 **Athena**	**WORKSHOP 1** ***Optimising Diagnostic Value in SPECT Myocardial Perfusion Imaging*** ***Chairs:*** *Ourania Demetriadou (Cyprus), Marta Wasilewska-Radwanska (Poland)*	
10:30 – 10:50	**W1-4** Development of the dynamic heart phantom control system	Andrzej Rosiek, Poland
10:50 – 11:10	**W1-5** Measurements and calculations of dose to the patient from Myocardial Perfusion Imaging studies	Demetris Kaolis, Cyprus
11:10 – 11:30	**W1-6** PEONY: software for image analysis in Nuclear Medicine	Charalambos Yiannakkaras, Cyprus
11:30 – 11:50	**W1-7** Monte Carlo Simulations to assess differentiation between defects in cardiac SPECT	Isabelle Chrysanthou, Cyprus
11:50 – 12:00	**Break**	

This Conference has received funding from the Cyprus Research Promotion Foundation and the European Regional Development Fund through the project ΥΓΕΙΑ/ΔΥΓΕΙΑ/0308/11: Optimising Diagnostic Value in SPECT Myocardial Perfusion Imaging.

ΕΥΡΩΠΑΪΚΗ ΕΝΩΣΗ

ΚΥΠΡΙΑΚΗ ΔΗΜΟΚΡΑΤΙΑ

ΔΙΑΡΘΡΩΤΙΚΑ ΤΑΜΕΙΑ

ΔΕΣΜΗ 2008

FRIDAY 25th of March 2011		
12:00 – 13:00 Athena	WORKSHOP 2 ***Beware of Pitfalls in Functional Imaging*** ***Chairs:*** *Chariklia Giannopoulou (Greece), Charalambos Yiannakkaras (Cyprus)*	
12:00 – 12:10	Introduction	Chariklia Giannopoulou, Greece
12:10 – 12:30	**W2-1** Pitfalls in Classical Nuclear Medicine 1	Christina Franghaki, Greece
12:30 – 13:00	**W2-2** Pitfalls in Classical Nuclear Medicine 2	Demitris Apostolopoulos, Greece
13:00 – 14:30	**Lunch**	
14:30 – 15:30 Athena	WORKSHOP 2 ***Beware of Pitfalls in Functional Imaging*** ***Chairs:*** *Chariklia Giannopoulou (Greece), Charalambos Yiannakkaras (Cyprus)*	
14:30– 14:50	**W2-3** Pitfalls in PET/CT 1	Nick Papathanassiou, Greece
14:50 – 15:10	**W2-4** Pitfalls in PET/CT 2	Phoivi Rondogianni, Greece
15:10 – 15:20	Discussion	
15:20 – 15:30	Conclusions	Chariklia Giannopoulou, Greece
15:30 – 16:00	**Break**	
16:00 – 18:30 Athena	WORKSHOP 3 ***Monte Carlo Simulations*** ***Chairs:*** *Michael Ljungberg (Sweden), Isabelle Chrysanthou (Cyprus)*	
16:00 – 17:00	**IL-3** Monte Carlo simulations for therapy imaging	Michael Ljungberg, Sweden
17:00 – 18:30	**HW1** Hands on Monte Carlo simulations	Michael Ljungberg, Sweden

19:30 – 23:00	**Conference Dinner & Dance**

This Conference has received funding from the Cyprus Research Promotion Foundation and the European Regional Development Fund through the project ΥΓΕΙΑ/ΔΥΓΕΙΑ/0308/11: Optimising Diagnostic Value in SPECT Myocardial Perfusion Imaging.

ΕΥΡΩΠΑΪΚΗ ΕΝΩΣΗ

ΚΥΠΡΙΑΚΗ ΔΗΜΟΚΡΑΤΙΑ

ΔΙΑΡΘΡΩΤΙΚΑ ΤΑΜΕΙΑ

ΔΕΣΜΗ 2008

FREDERICK RESEARCH CENTRE

AGH

IBE

SATURDAY 26th of March 2011		
08:00 – 10:10 **Athena**	SCIENTIFIC SESSION 4 ***Simulations*** ***Chairs:*** *Isabelle Chrysanthou (Cyprus), Marta Wasilewska-Radwanska (Poland)*	
08:00 –08:30	**KN-2** Dynamic Phantoms for Nuclear Medicine Studies	Marta Wasilewska-Radwanska, Poland
08:30 – 08:50	**SS4-1** Initial tests of a new phantom for investigation of spatial resolution, partial volume effect and detectability in nuclear medicine tomography	Marcus Söderberg, Sweden
08:50 – 09:10	**SS4-2** Design of heart defects for severity simulation	Isabelle Chrysanthou, Cyprus
09:10 – 09:30	**SS4-3** Fast Analytic Simulation Toolkit for Generation of 4D PET-MR Data from Real Dynamic MR Acquisitions	Charalampos Tsoumpas, UK
09:30 – 09:50	**SS4-4** Study of the effect of magnetic field in positron range using GATE simulation toolkit	George Soultanidis, Greece
09:50 – 10:10	**SS4-5** Scatter Correction in 3D PET using STIR	Irene Polycarpou, UK
10:10 – 10:30	**Coffee Break**	
10:30 – 11:50 **Athena**	SCIENTIFIC SESSION 5 ***Clinical and Quality Management*** ***Chairs:*** *Stella Koumna (Greece/Cyprus), Prodromos Kaplanis (Cyprus)*	
10:30 –10:50	**SS5-1** NEMA NU 1-2007 based and independent Quality Control software for gamma and SPECT cameras	Troels Jorgensen & Robin de Nijs, Denmark
10:50 – 11:10	**SS5-2** Implementing a Quality Management System in a Nuclear Medicine Department	Demetris Kaolis, Cyprus
11:10 – 11:30	**SS5-3** Specificity and Sensitivity of SPECT Myocardial Perfusion Studies at the Nuclear Medicine Department of Limassol General Hospital in Cyprus	Stella Koumna, Greece/Cyprus
11:30 – 11:50	**SS5-4** Scintigraphic Estimation of the Relative Liver Perfusion in the Choice of Treatment of Liver Carcinomas	Mirjana Stojkovic, Serbia
11:50 – 12:00	**Break**	

This Conference has received funding from the Cyprus Research Promotion Foundation and the European Regional Development Fund through the project ΥΓΕΙΑ/ΔΥΓΕΙΑ/0308/11: Optimising Diagnostic Value in SPECT Myocardial Perfusion Imaging.

ΕΥΡΩΠΑΪΚΗ ΕΝΩΣΗ

ΚΥΠΡΙΑΚΗ ΔΗΜΟΚΡΑΤΙΑ

ΔΙΑΡΘΡΩΤΙΚΑ ΤΑΜΕΙΑ

ΔΕΣΜΗ 2008

FREDERICK RESEARCH CENTRE

AGH UNIVERSITY OF SCIENCE AND TECHNOLOGY KRAKOW POLAND

IBE

SATURDAY 26th of March 2011		
12:00 – 13:30 **Athena**	**EFOMP SESSION** ***Education and Training of the Medical Physicist in Nuclear Medicine*** ***Chairs:*** *Carmel J. Caruana (Malta), Glenn Flux (United Kingdom)*	
12:00 – 12:20	**ES1** The Activities of EFOMP in Education and Training: Past, Present and Future	Stelios Christofides, Cyprus
12:20 – 12:40	**ES2** EC Project `GUIDELINES ON MPE´: Proposed Qualification and Curriculum Frameworks, and the MPE in Nuclear Medicine	Carmel Caruana, Malta
12:40 – 13:00	**ES3** EANM Education and Training	Glenn Flux, UK
13:00 – 13:20	**ES4** Multi-type content repurposing and sharing in medical education, The mEducator project	Athos Antoniades, Cyprus
13:20 – 13:30	Discussion	
13:30 – 14:00 **Athena**	**CLOSING CEREMONY** ***Chairs:*** *Stelios Christofides (Cyprus), Yiannis Parpottas (Cyprus)*	
13:00 – 13:30	Farewell and Closing Remarks	

This Conference has received funding from the Cyprus Research Promotion Foundation and the European Regional Development Fund through the project ΥΓΕΙΑ/ΔΥΓΕΙΑ/0308/11: Optimising Diagnostic Value in SPECT Myocardial Perfusion Imaging.

ΕΥΡΩΠΑΪΚΗ ΕΝΩΣΗ

ΚΥΠΡΙΑΚΗ ΔΗΜΟΚΡΑΤΙΑ

ΔΙΑΡΘΡΩΤΙΚΑ ΤΑΜΕΙΑ

ΔΕΣΜΗ 2008

Journal of Physics: Conference Series **317** (2011) 012001 doi:10.1088/1742-6596/317/1/012001

Evaluation of two approaches to motion-corrected PET image reconstruction

I Polycarpou, P K Marsden and C Tsoumpas

Division of Imaging Sciences & Biomedical Engineering, School of Medicine, King's College London, UK

E-mail: irene.polycarpou@kcl.ac.uk

Abstract. Motion correction is a major task in PET. This study aims to understand the behaviour of different approaches to motion-corrected image reconstruction in terms of convergence rates and the properties of the images obtained. We have studied two approaches that have been suggested to motion-correct PET gated data. The first method, RRA, is based on independent reconstructions of each gate which are then registered to a reference gate and averaged. On the other hand, in the second method, MCIR, all data are used with the motion information within a single reconstruction. As a first step, this investigation compares the methods with no motion present. Multiple simulations have been produced with counts approximating a 5 min PET thorax acquisition. The image bias and variance are the main figures of merit. The results show different convergence and statistics performance between the two methods. RRA converges faster than MCIR but not at the correct value. Considering the trade-off between bias and variance, it is seen that MCIR outperforms RRA. It has been also demonstrated that, although at higher iterations RRA has lower variance and higher bias than MCIR, at a low number of iterations the performance of the methods becomes closer. This investigation has revealed the statistical and computational discrepancies between RRA and MCIR when motion is not present and MCIR is shown to be clearly superior.

1. Introduction

Respiratory motion is an important source of resolution degradation and hence quantification, being a major challenge in PET research. A common method to overcome motion is gating, during which respiratory positions can be reconstructed producing a "motion-free" noisy image. For this reason it has been proposed to motion-correct by estimating motion between gates, thus using all the measured data for the final image. The most common approaches to motion-correction are the Reconstruct-Register-Average (RRA) and Motion-Compensated Image Reconstruction (MCIR).

The RRA is based on independent reconstructions of each position which are then registered to a reference position and averaged [1-2]. During this method, the final image is obtained post-reconstruction without taking into account the non-linearity effects of OSEM reconstruction algorithm which are exaggerated due to the relatively low counts in each gate.

Several investigators proposed to modify the reconstruction algorithms by incorporating motion information into the system model, referred to as MCIR [3]. Considering all data allows the total emission statistics to contribute to the final image. However, modifying the system model makes the reconstruction process computationally more demanding and might also affect convergence properties.

Although, there has been a comparative study [2], there is still not enough evidence that clearly supports which method is practically better for PET respiratory motion correction. Comparison of the methods requires assessment in terms of noise and bias. This investigation reveals the statistical discrepancies between RRA and MCIR when motion is not present providing a complete characterization of the source and extent of the differences without the additional complications of how motion is incorporated.

2. Methods

The RRA and MCIR methods have been compared without motion. Reconstructions were performed with the Software library for Tomographic Image Reconstruction (STIR) [4]. A procedure to simulate realistic clinical FDG-18 PET simulation data was used [5]. The PET data were forward projected simulating a 2D-acquisition of the Philips Gemini scanner. Following forward projection, attenuation and Poisson noise were added such that the overall acquisition corresponds to a 5 min 2D PET thorax acquisition, but with 3D noise levels (i.e. 5×10^7 counts). One thousand statistically independent realizations of the acquisition were simulated. For MCIR, each realization was reconstructed, whereas for RRA, each replication corresponded to ten sets of projection data preserving Poisson statistics. FBP (with ramp filter) and OSEM (23 subsets) reconstruction algorithms were performed. Each slice consists of 128×128 pixels with size 4 mm each, and the entire volume consists of 87 slices with 2 mm thickness. The comparison of the two methods is assessed by computing voxel-wise figures of merits using the reconstructed images of the 1000 realizations. The following statistical information is presented: bias, standard deviation, and variance.

3. Results

Figure 1 shows the transverse planes of the bias images obtained from 1000 realizations using the RRA and MCIR methods with the FBP and OSEM algorithms (40 iterations with 23 subsets). The reconstructed images are 3D and thus only one transverse plane (i.e. 67) is illustrated. Figure 2 presents the plots of the percentage bias (%) for different regions of interest (ROIs) as a function of the iteration number for OSEM, whereas lines are plotted for the FBP algorithm cases.

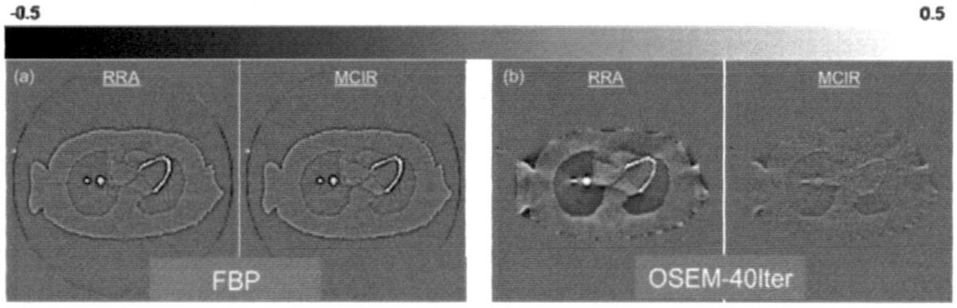

Figure 1. Bias images from 1000 realizations using the RRA and MCIR methods with the algorithms: (a) FBP, and (b) OSEM (40 iterations).

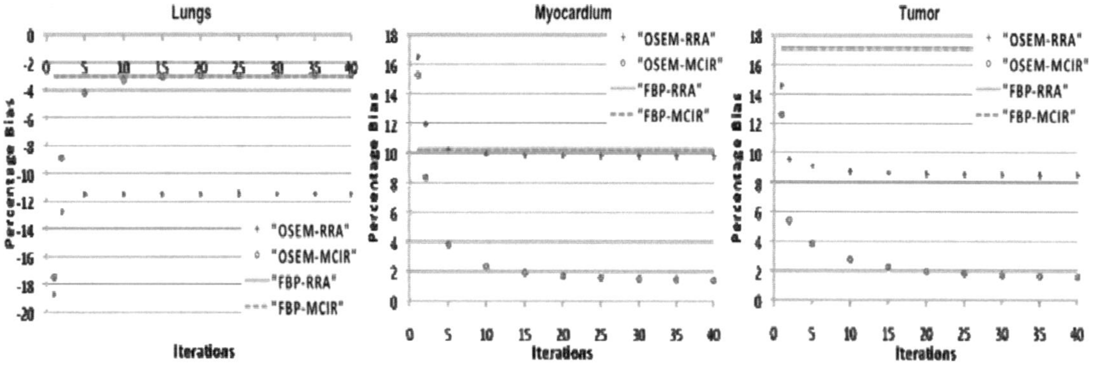

Figure 2. Plots of percentage bias as a function of iteration number using the RRA and MCIR methods and, the OSEM and the FBP algorithms for (a) lungs, (b) myocardium, (c) tumor.

Figure 3 illustrates the transverse planes of the variance images obtained from 1000 realizations using the methods RRA and MCIR for the FBP and OSEM (40 iterations with 23 subsets) algorithms for. In figure 1, only the transverse plane 67 is illustrated.

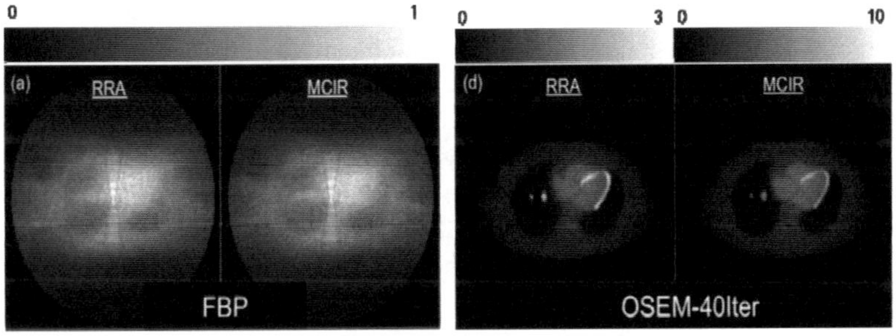

Figure 3. Variance images from 1000 realizations using RRA and MCIR with the algorithms: (a) FBP, and (b) OSEM (40 iter.).

Figure 4 presents the plots of the bias in respect to the standard deviation for the OSEM algorithm using different number of iterations. Different ROIs are presented.

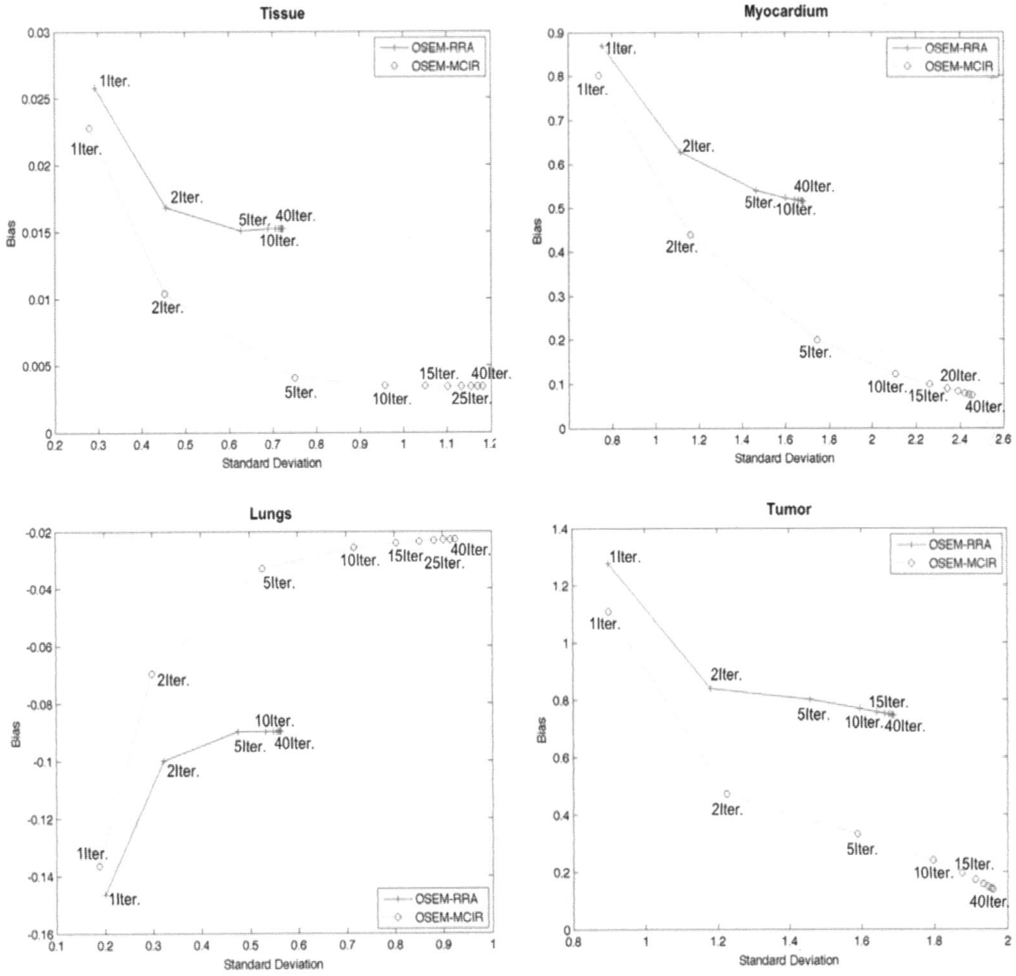

Figure 4. Plots of bias versus standard deviation for RRA and MCIR using OSEM (1 to 40 it.) for (a) tissue, (b) myocardium, (c) lungs, (d) tumor.

4. Discussion

The FBP results of figures 1 and 2 point out low and equal bias for RRA and MCIR in areas and edges which can be attributed to reconstruction artefacts. Substantial differences are observed between RRA and MCIR using OSEM; overestimations and underestimations in low and high counts areas respectively. For all regions, higher bias is observed for RRA than MCIR while for both methods the bias is progressively decreased as the number of iterations is increased. In figure 2, the percentage bias after a high number of iterations in lungs was -11 % for RRA, and -3 % for MCIR whereas in tumour was 10 % for RRA, and 2 % for MCIR.

Figure 3 demonstrates that the variance of the FBP for both RRA and MCIR is identical and fairly uniform for all regions. Comparing RRA and MCIR using the OSEM algorithm in figure 3, it can be clearly seen that the two methods have similar variance for a low number of iterations; whereas when the number of iterations is increased, the difference between the two methods becomes larger, in particular for MCIR.

Figure 4 verifies the trade-off between bias and variance. RRA increased faster than MCIR, especially for low or high count areas. It is also clearly demonstrated that although at higher iterations RRA has lower variance and higher bias than MCIR, at low number of iterations (i.e. 1 or 2), which currently is the typical setting for clinical PET, the performance of the two methods becomes closer. The MCIR has lower bias than RRA at matched levels of variance.

5. Conclusions

This investigation has revealed the statistical and computational discrepancies between RRA and MCIR when motion is not present. The main finding is the substantially different convergence and statistical performance between them. RRA causes high levels of overestimation and underestimation in cold and hot regions respectively. It is found that even for a small iteration number (e.g. 2), MCIR has lower bias and similar variance. The trade-off between bias and variance for both methods shows that MCIR outperforms RRA. The next step is the integration of motion correction and the study of its impact.

Acknowledgments

I. Polycarpou work is funded by a King's College London studentship and C. Tsoumpas work has been supported by the European Union under the 7th framework program (No. 201651).

References

[1] Dawood M, Buther F, Jiang X and Schafers K P 2008 Respiratory motion correction in 3-D PET data with advanced optical flow algorithms *IEEE Trans. Med. Imag.* **27** 1164-75

[2] Thielemans K, Manjeshwar R M, Xiaodong T and Asma E 2007 Lesion detectability in motioncompensated image reconstruction of respiratory gated PET/CT *IEEE Nucl. Sci. Symp. Med. Imag. Conf. Rec.* **6** 3278-82

[3] Jacobson M W and Fessier J A 2006 Joint estimation of respiratory motion and activity in 4D PET using CT side information *IEEE Inter. Symp.: Biom. Imag.* 275-78

[4] Thielemans K, Mustafovic S and Tsoumpas C 2006 STIR: Software for Tomographic Image Reconstruction Release 2 *IEEE Nucl. Sci. Symp. Med. Imag. Conf.* **4** 2174-76

[5] Tsoumpas C, Buerger C, King A P, Keereman V, Vandenberghe S, Schulz V, Schaeffter T and Marsden P K 2009 Simulation of dynamic PET data from real MR acquisitions *IEEE Nucl. Sci. Symp. Med. Imag. Conf. Rec.* 3065-68

International Conference on Image Optimisation in Nuclear Medicine (OptiNM) IOP Publishing
Journal of Physics: Conference Series **317** (2011) 012002 doi:10.1088/1742-6596/317/1/012002

Reconstruction of linear kinetic parameters directly from projection PET data

G I Angelis[1,2], A C Tziortzi[3,4] and C Tsoumpas[1,5]

[1] MRC Clinical Sciences Centre, Imperial College London, Hammersmith Hospital Campus, Du Cane Road, W12 0NN, UK
[2] Department of Cancer and Enabling Sciences, MAHSC, Wolfson Molecular Imaging Centre, The University of Manchester, Manchester, M20 3LJ, UK
[3] FMRIB Centre, Department of Clinical Neurology, University of Oxford, UK
[4] GlaxoSmithKline, Clinical Imaging Centre, Imperial College London, UK
[5] Division of Imaging Sciences & Biomedical Engineering, School of Medicine, King's College London, UK

E-mail: georgios.angelis@mmic.man.ac.uk

Abstract. Dynamic Positron Emission Tomography (PET) data provide functional information. Usually, this is measured in the form of pharmacokinetic parameters derived from the temporal response of each region. Recent trends have shown that when pharmacokinetic parameters are estimated directly from the projection data, they are less affected by noise. This work investigates an existing parametric maximum likelihood expectation maximization algorithm applied to [18F]DOPA data using reference-tissue input function. The study reveals how direct reconstruction of pharmacokinetic parameters from the measured data can be performed optimally. It explains how to optimize the speed of the standard iterative algorithm and it compares the results with the existing FBP method. The improvement of the quality of the parametric images preserving quantification suggests the usefulness of direct estimation of the kinetic parameters. This algorithm is freely available within the open-source library STIR 2.1.

1. Introduction

Dynamic PET data can provide physiologically meaningful parameters about the biological process of interest. This is done by independently reconstructing all dynamic frames and then applying an appropriate kinetic model to the time-activity curve (TAC) of a region of interest (ROI). However, it is desirable to measure parameters at very small regions (e.g. voxels) in order to detect functional heterogeneity. Nevertheless, it is difficult to perform voxel-wise parametric images since they tend to be noisy. Additionally, the kinetic modelling requires appropriate weights in order to model the noise distribution in the reconstructed images, which is very difficult, because noise depends on multitude of parameters [1].

However, these limitations can be eliminated when the kinetic model is incorporated within the reconstruction algorithm and the kinetic parameters are estimated directly from the measured data. In this case, the noise distribution in the measured data can be accurately modelled, since it is well approximated by Poisson statistics, leading to statistically improved parameter estimation [2].

Recently, it has been clearly demonstrated that particularly for FDG brain studies, direct parametric reconstruction, using the Parametric Ordered Subsets Expectation Maximization (POSEM) algorithm, outperforms conventional post reconstruction methods using standard Patlak analysis [3]. However, as pointed out by Tsoumpas *et al.* [4], POSEM usually demonstrates impractically slow convergence rates.

In this paper, we apply POSEM to [^{18}F]DOPA brain studies to estimate the tracer uptake in the striatal region and attempt to optimize the convergence rate. In order to do that, we employed a progressively decreasing subset scheme during the reconstruction, which allowed faster and more accurate parameter estimation compared to the original POSEM algorithm.

2. Materials and methods

The reference-region Patlak plot [5] was applied to clinical datasets related to striatal dopaminergic activity studies. The data were acquired using the Siemens ECAT® EXACT™ 3D, which operates in fully 3D acquisition mode storing all measured PET data in list-mode. List-mode data were rebinned into sinograms according to the scan protocol. The bilinear Patlak plot was incorporated within an ordinary Poisson algorithm, allowing the acquisition of two parametric images representing the tracer influx in the tissue of interest and the tracer free fraction. The reference tissue TAC was extracted from cerebellum using a ligand specific template image and an ad-hoc atlas.

The open-source software package STIR (release 2.1) [6] was used to process PET data and reconstruct dynamic and parametric images. The post reconstruction parametric maps were acquired in two steps: First, the dynamic image was reconstructed using FBP-3DRP algorithm, and then the Patlak plot was applied to the last 8 frames (5 minutes each).

For POSEM, three different subset schemes were investigated: (a) "6-6-1": 6 subsets for 460 iterations and then 260 iterations of PMLEM; (b) "18-6-1": 18 subsets for 230 iterations then 6 subsets for another 230 iterations and finally 260 PMLEM iterations; and (c) "18-1-1": 18 subsets for 230 iterations and then 520 PMLEM iterations. All reconstruction schemes were initialized by the post reconstruction FBP image and let run approximately for 750 iterations. The different multi-subsets schemes were evaluated in terms of convergence rate of the mean ROI value on striatum which were automatically delineated using the CIC atlas [7]. The post reconstruction parametric maps, obtained with POSEM were quantitatively compared to FBP using a voxel-wise Bland-Altman for all voxels located in the area of interest (i.e. striatum). To reduce noise for this specific comparison, both parametric images were post-filtered with a 3 mm FWHM Gaussian kernel.

Table 1. Mean ROI and standard deviation values on striatum for the final Patlak slope images of each POSEM scheme and the FBP image.

Uptake (10^{-4} s^{-1})	Patient		Normal	
	Mean	Std. Dev	Mean	Std. Dev
FBP	2.21	± 1.70	1.76	± 1.42
POSEM (6-6-1)	2.06	± 1.86	1.79	± 1.61
POSEM (18-6-1)	2.07	± 1.97	1.78	± 1.70
POSEM (18-1-1)	2.07	± 1.94	1.79	± 1.67

3. Results

A comparison among the descending-subset schemes under investigation is presented in table 1 and figure 1. All schemes appear to converge to the same ROI value for both parameters, but those employing more subsets (i.e. 18) converge faster. A quantitative comparison between FBP and POSEM (scheme 18-6-1) for all voxels in striatum of a normal study is presented in figure 2 using the Bland-Altman plot for both parameters. Both cases indicate better visual quality for POSEM compared to the indirect FBP. Finally, in figure 3 a qualitative comparison of the Patlak slope sagittal and transaxial views of the striatum between the indirect FBP and POSEM is presented for one normal volunteer and one patient.

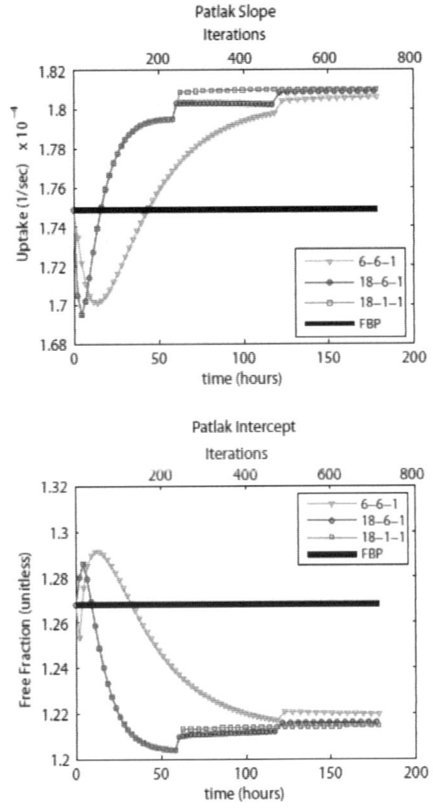

Figure 1. Mean ROI values for the slope (up); and intercept (bottom) on striatum with schemes: 6-6-1, 18-6-1 and 18-1-1.

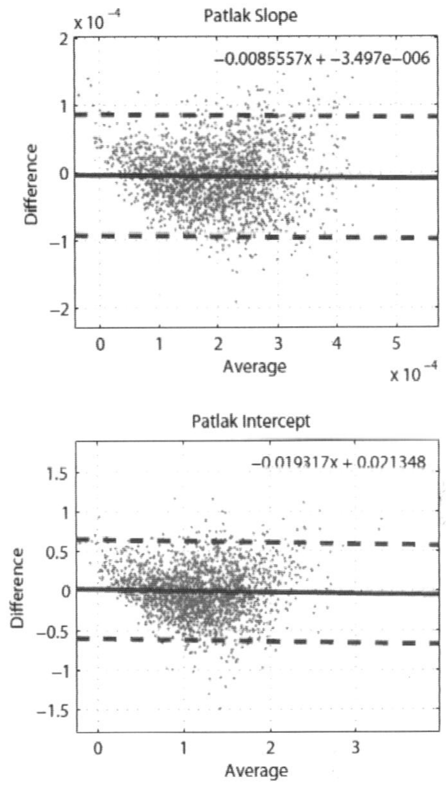

Figure 2. Voxel-wise Bland Altman plots comparing the voxel values in striatum for the slope (up) and intercept (bottom) parameters.

4. Discussion

The results obtained from different subset combinations showed that there is no practical difference in the final estimated values (figure 1). However, schemes started with more subsets showed higher speed-up compared to those started with smaller subsets. Therefore, it is recommended to start the reconstruction using a large number of subsets, in order to take advantage of the fast convergence rate and then switch to a smaller number of subsets, or even to PMLEM in an attempt to avoid the limit subset cycle and achieve better accuracy in the finally estimated parameter. However, there is trade-off between faster convergence and increased standard deviation in the reconstructed images (table 1).

Voxel-wise Bland-Altman plots were employed in order to investigate the quantitative agreement between the parametric maps of the proposed method (i.e. POSEM; scheme 18-6-1) and the well-established post-reconstruction FBP method (figure 2). The results showed good quantitative agreement between the two methods for both parameters (i.e. slope and intercept), since the linear regression line (black line) is almost horizontal, crossing the zero of the vertical axis. However, bias is observed for low parametric values due to the non-negativity constraint induced by POSEM. This bias is high in low-uptake regions, e.g. cerebellum, where the effect of non-negativity is prominent.

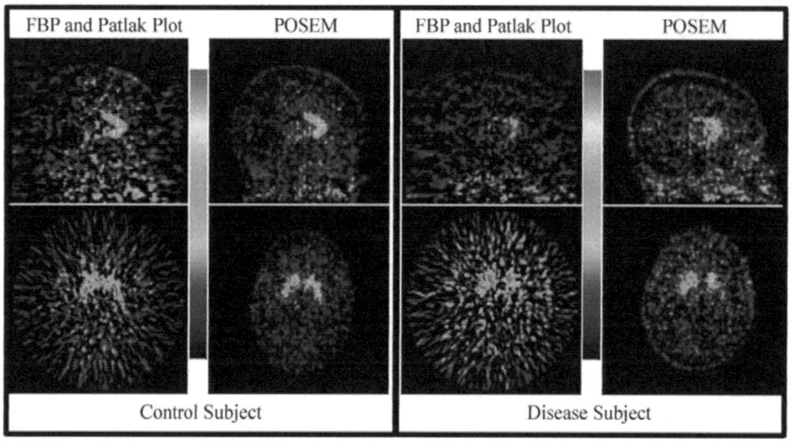

Figure 3. Comparison of Patlak slope images for patient and control subjects. The FBP image was used to initialize POSEM. The direct image is the result of the subset scheme 18-6-1.

Finally, visual inspection of the reconstructed images (figure 3) indicates that POSEM produces parametric images of better quality than the conventional indirect FBP. Nevertheless, the images are still noisy. Regularization approaches are expected to be advantageous and may also reduce bias if the penalty terms are carefully selected [8].

5. Conclusions

The results of this work indicate that POSEM is applicable to [^{18}F]DOPA brain studies, producing images with better visual quality compared to the conventional indirect methods and preserving quantification. Convergence has been optimized by gradual reduction of the number of subsets over iterations, though the computational demand is still very high [9].

Acknowledgments

Written informed consent was obtained from all subjects. The authors are thankful to Dr K Thielemans and Dr F E Turkeimer for the scientific advice throughout the project, as well as Mr Simon Burbidge for the technical advice and support using the Imperial College London HPC system: http://www.imperial.ac.uk/ict/services/teachingandresearchservices/highperformancecomputing

References

[1] Wilson D W, Tsui B M W and Barrett H H 1994 *Phys. Med. Biol.* **39** 847-71
[2] Matthews J C, Angelis G I, Kotasidis F A, Markiewicz P J and Reader A J 2010 *IEEE Nucl. Sci. Symp.Conf. Rec.* **M10-6**
[3] Tsoumpas C, Turkheimer F E and Thielemans K 2008 *Med. Phys.* **35** 1299-309
[4] Tsoumpas C, Turkheimer F E and Thielemans K 2008 *Med. Phys.* **35** 3963-71
[5] Patlak C S, Blasberg R G and Fenstermacher J D 1983 *J. Cerebr. Blood Flow Metab.* **3** 1-7
[6] Thielemans K, Mustafovic S and Tsoumpas C 2007 *IEEE Nucl. Sci. Symp. Conf. Rec.* **4** 2174-76
[7] Tziortzi A C, Searle G E, Tzimopoulou S, Salinas C, Beaver J D, Jenkinson M, Laruelle M, Rabiner E A and Gunn R N 2011 *NeuroImage* **54** 264-277
[8] Wang G, Fu L and Qi J 2008 *Phys. Med. Biol.* **53** 593-604
[9] Angelis G I, Thielemans K, Tziortzi A C, Turkheimer F E and Tsoumpas C 2011 *Comp. Med. Imaging Graph* (in press)

Journal of Physics: Conference Series **317** (2011) 012003 doi:10.1088/1742-6596/317/1/012003

Justification of administered dose level in brain perfusion imaging with ^{99m}Tc-HMPAO

A P Stefanoyiannis[1], I Gerogiannis[2], X Geronikola-Trapali[1], I Armeniakos[1], A Prentakis[1], S Soultanis[3] and S N Chatziioannou[1]

[1] University General Hospital of Athens "Attikon", 1 Rimini Street, GR-12462, Athens, Greece
[2] Nicosia General Hospital, 215 Old Nicosia-Limassol Road, CY-2029, Nicosia, Cyprus
[3] Konstantopouleion Hospital "Agia Olga", 3-5 Agias Olgas Street, GR-14233, Athens, Greece

E-mail: apstefano@yahoo.gr

Abstract. Brain perfusion imaging by means of 99mTc-HMPAO is widely used in the diagnosis of Alzheimer's disease. The administered dose range recommended by the manufacturer and reported in bibliography is rather wide (~ 9.5 - 27 mCi), necessitating further quantitative analysis. In the framework of this study, a quantitative evaluation of the radiopharmaceutical performance for different values of administered dose was carried out, based on image quality indicators. Evaluation of image quality was based on wavelet-generated contrast, noise, and contrast-to-noise ratio indicators, denoted as CI, NI and CNR respectively. Subsequently, a generic image quality index was correlated with the administered dose, to produce an overall performance indicator (denoted as PI). Application of appropriate statistical tests (analysis of variance for normal and Kruskal-Wallis test for non-normal distributions) showed that there is a statistically significant difference in CI ($p < 0.01$), NI ($p < 0.001$) and CNR ($p < 0.05$), but not in PI ($p > 0.05$) values. Application of Tukey test for CI and NI normal distributions demonstrated that CI (10 mCi) = CI (20 mCi) < CI (15 mCi) and NI (10 mCi) > NI (20 mCi), while NI (15 mCi) could not be characterised. Finally, application of non-parametric multiple comparisons showed that CNR (20 mCi) > CNR (10 mCi), while CNR (15 mCi) could not be characterised. Consequently, brain perfusion imaging, by means of 99mTc-HMPAO utilising an administered dose of 20 mCi, results in improved image quality on the basis of the estimated indicators. Additionally, this image quality improvement is sufficient to justify the increased patient radiation burden.

1. Introduction

Modern clinical practice is based on a plethora of diagnostic tools for the accurate diagnostic assessment of various forms of brain pathology. Brain perfusion imaging by means of ^{99m}Tc-labeled hexamethyl propylene amine oxime (HMPAO) is a well-established Nuclear Medicine diagnostic procedure [1]. It is a method of great clinical importance for the diagnosis of Alzheimer's disease, through the differentiation of small changes in the regional cerebral blood flow [2, 3]. Recent studies performing patient-by-patient comparison between this technique and DSC-MRI (a technique based on magnetic resonance imaging) proved the superiority of the former in the diagnosis of Alzheimer's disease, as well as the other forms of dementia [1, 4].The administered dose range recommended by the supplying company and reported in litterature is rather wide (approximately 9.5 - 27 mCi). This fact necessitates further quantitative analysis of the technique, so as to minimise patient absorbed dose without compromising the examination diagnostic value. In this study, a quantitative evaluation of the radiopharmaceutical performance for different values of administered dose (10, 15, 20 mCi) was performed, via application of image analysis techniques [5]. Through this cost-to-benefit type analysis, the justification of higher administered dose levels in order to perform the specific diagnostic procedure was examined.

2. Materials and methods

A sample of 78 patients (classified as normal or presenting various forms of pathology) was utilised. The administered dose was 10 mCi for 56 patients, 15 mCi for 10 patients and 20 mCi for 12 patients. Evaluation of image quality was based on contrast, noise and contrast-to-noise ratio indicators, denoted as CI, NI and CNR respectively. Calculation of all indicators was based on wavelet transform [6-10]. An overall performance indicator (denoted as PI) was also utilised. Its calculation was based on the ratio of CNR, which was considered as a generic image quality index, by the administered dose.

3. Results

Figure 1 shows the results of the estimation of image quality indicators CI, NI and CNR and the overall performance indicator PI for the subsamples of 10, 15 and 20 mCi.

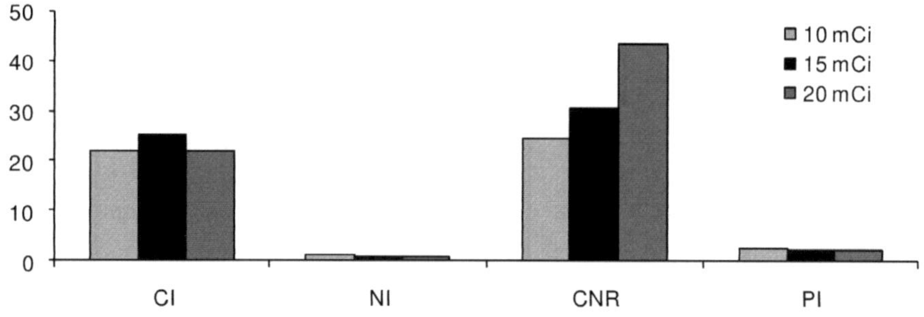

Figure 1. Estimation of the image quality indicators CI, NI and CNR, and the overall performance indicator PI for the subsamples of 10, 15 and 20 mCi.

The results of quantitative evaluation are presented in table 1, where the % differences of the estimated indicators for 15 mCi and 20 mCi subsamples versus the control subsample of 10 mCi are reported.

Table 1. The % differences of the estimated indicators for 15 mCi and 20 mCi subsamples in respect to (vs.) the control subsample of 10 mCi.

	CI	NI	CNR	PI
% diff. (15 mCi vs. 10 mCi)	14.58	-25.13	24.41	-17.06
% diff. (20 mCi vs. 10 mCi)	-1.18	-52.09	78.01	-11.00

Table 2 shows the results of further statistical analysis. Calculation of skewness parameter revealed the normality of CI and NI, and the non-normality of CNR and PI distributions. Application of appropriate statistical tests (analysis of variance for normal and Kruskal-Wallis test for non-normal distributions) showed that there is a statistically significant difference in CI ($p < 0.01$), NI ($p < 0.001$) and CNR ($p < 0.05$), but not in PI ($p > 0.05$) values. Application of Tukey test for normal distributions of CI and NI demonstrated that CI (10 mCi) = CI (20 mCi) < CI (15 mCi) and NI (10 mCi) > NI (20 mCi), while NI (15 mCi) could not be characterised. Finally, application of non-parametric multiple comparisons showed that CNR (20 mCi) > CNR (10 mCi), while CNR (15 mCi) could not be characterised.

Table 2. Statistical results for the image quality indicators.

Parameter	Distribution	Statistical test	Result
CI	normal	ANOVA	SS ($p<0.01$)
NI	normal	ANOVA	SS ($p<0.001$)
CNR	non-Gaussian	Kruskal-Wallis	SS ($p<0.05$)
PI	non-Gaussian	Kruskal-Wallis	SNS ($p>0.05$)

4. Conclusions

According to the results of the present study, brain perfusion imaging by means of ^{99m}Tc- HMPAO, utilising the administered dose values examined (20 mCi), results in improved image quality on the basis of the estimated indicators. Additionally, this image quality improvement is sufficient to justify the increased patient absorbed dose.

As an open issue, the future study of image quality for higher values of administered dose is mentioned, always in the framework of internationally acceptable clinical practice. The comparison among ^{99m}Tc-HMPAO scintigraphic images, produced by different administered dose values, could be based on image quality indicators, in addition to a clinical evaluation of brain features crucial for the diagnosis.

References

[1] Cavallin L, Danielsson R, Oksengard R, Wahlund L O, Julin P, Frank A, Engman E L, Svensson L and Kristoffersen Wiberg M 2006 Can dynamic susceptibility magnetic resonance imaging replace single-photon emission computed tomography in the diagnosis of patients with Alzheimer's disease? A pilot study *Acta Radiol.* **9** 977-85

[2] Ansar B, Osaki Y, Hiroaki K, Oku N, Takasawa M, Kimura Y, Begum N N, Ikejiri Y, Takeda M and Hatazawa J 2006 Effect of linearization correction on statistical parametric mapping: a 99mTc-HMPAO brain perfusion SPECT study in mild Alzheimer's disease *Ann. Nucl. Med.* **20**(8) 511-17

[3] Koulibaly P M, Nobili F, Migneco O, Vitali P, Robert P H and Girtler N 2003 99mTc-HMPAO and 99mTc-ECD perform differently in typically hypoperfused areas in Alzheimer's disease *Eur. J. Nucl. Med. Mol. Imaging* **30** 1009-13

[4] Kubota T, Ushijima H, Yamada K, Okuyama O, Nishimura T 2005 Diagnosis of Alzheimer's disease using brain perfusion SPECT and MR imaging: which modality achieves better diagnostic accuracy? *Eur. J. Nucl. Med. Mol. Imaging* **32** 414-421

[5] Stefanoyiannis A P, Geronikola-Trapali X, Armeniakos I, Chatziioannou S N, Prentakis A, Efstathopoulos E and Gouliamos A 2007 Effect of administered dose level on image quality of brain perfusion imaging with ^{99m}Tc-HMPAO: a quantitative analysis *Eur. J. Nucl. Med.* **34** suppl. 2 S340

[6] Laine A, Fan J and Yang 1995 Wavelets for contrast enhancement of digital mammography *IEEE Eng. Med. Biol.* 536-51

[7] Mallat S and Hwang W 1992 Singularity detection and processing with wavelets *IEEE Trans. Inform. Theory* **38** 617-43

[8] Zong X and Laine A 1996 *Proc. of SPIE, Wavelet Applications III* **2762** 566-574

[9] Costaridou L, Sakellaropoulos P, Stefanoyiannis A P, Ungureanu E and Panayiotakis G 2001 *Br. J. Radiol.* **74** 913-19

[10] Stefanoyiannis A P, Costaridou L, Skiadopoulos S and Panayiotakis G 2003 *Eur. J. Radiol.* **45** 139-49

International Conference on Image Optimisation in Nuclear Medicine (OptiNM) IOP Publishing
Journal of Physics: Conference Series **317** (2011) 012004 doi:10.1088/1742-6596/317/1/012004

A comparison between the conventional manual ROI method and an automatic algorithm for semiquantitative analysis of SPECT studies

L Pagan[1], C Tranfaglia[2], S Galli[2], G Lucchi[2], B Novi[1], G Fagioli[2], G Guidarelli[1]

[1] Health Physics Unit, Maggiore Hospital of Bologna, Italy
[2] Nuclear Medicine Unit, Maggiore Hospital of Bologna, Italy

E-mail: laura.pagan@ausl.bo.it

Abstract. In this study, the performance of a free software for automatic segmentation of striatal SPECT brain studies (BasGanV2 – www.aimn.it) and a standard manual Region Of Interest (ROI) method were compared. The anthropomorphic Alderson RSD phantom, filled with solutions at different concentration of ^{123}I-FP-CIT with Caudate-Putamen to Background ratios between 1 and 8.7 and Caudate to Putamen ratios between 1 and 2, was imaged on a Philips-Irix triple head gamma camera. Images were reconstructed using filtered back-projection and processed with both BasGanV2, that provides normalized striatal uptake values on volumetric anatomical ROIs, and a manual method, based on average counts per voxel in ROIs drawn in a three-slice section. Caudate-Putamen/Background and Caudate/Putamen ratios obtained with the two methods were compared with true experimental ratios. Good correlation was found for each method; BasGanV2, however, has higher R index (BasGan $R_{mean} = 0.95$, $p < 0.001$) than the ROI method ($R_{mean} = 0.89$, $p < 0.005$). The results obtained suggest that this new software can be suitable in daily practice, being an accurate, simple, fully automatic and reproducible method of semiquantitative analysis of ^{123}I-FP-CIT SPECT data with, moreover, the advantage of the availability of a control subject's database.

1. Introduction

It is well known that in the last few years SPECT studies of the pre-synaptic dopaminergic system has gained a central role in the study of parkinsonian syndromes [1], and particularly in differentiating Parkinsonism from Essential Tremor and Lew Bodies Dementia from Alzheimer disease [2,3].
Qualitative, visual analysis of striatal uptake is probably the most common approach and have demonstrated to yield similar sensitivity and specificity as a non-anatomical ROI method. The semiquantitative approach however can reach a higher positive predictive value and a better likelihood ratio [4].

Furthermore, the results derived from ROIs have potential advantages because they provide a continuous variable (rather than a categorical variable), which is more likely to detect subtle changes in DAT binding in striatal subregions and allow monitoring of disease progression [3,5].
Because of this, different methods of semiquantitative analysis have been proposed to reach a better accuracy in ROI definition and activity measurement than the standard manual method.

Recently a new version of Basal Ganglia software (BasGanV2) for semiquantitative analysis of SPECT studies with ^{123}I-FP-CIT (DatSCAN) is available on the website of Italian Nuclear Medicine

Association (www.aimn.it) [6]. BasGanV2 is easy to use, freely downloadable from the web and presents the advantage of automatic procedures: is user-independent and provides 3D information of the brain since it is based on a template derived from Talairach and Tournoux neuroanatomical atlas. Moreover the software tool incorporates a customisable database of controls patients.

This software tool can be used for automated and standardized processing, evaluation and documentation of DatSCAN SPECT studies, and has demonstrated its usefullness in the differential diagnosis between Parkinsonism and essential tremor [7].

The aim of this study was to evaluate the performance of both BasGanV2 and a standard method, based on manual ROI plotting, with the use of anthropomorphic striatal phantom images reconstructed by a filtered back-projection (FBP) algorithm.

2. Materials and Methods

Nine SPECT acquisitions of the Alderson phantom (designed ad hoc to reproduce the striatal morphology) were performed with the striatum and background chambers filled with solutions at different concentration of ^{123}I-FP-CIT (GE Healthcare, Amersham, UK) in order to obtain different specific caudate to background (C/B), putamen to background (P/B) and caudate to putamen (C/P) ratios. The C-P/B and C/P ratios obtained experimentally in the nine simulations ranged from 1 to 8.7 and 1 to 2 (see table 1).

Data acquisition were carried out with a Philips-Irix thriple head gamma camera equipped with low-energy, high resolution parallel hole collimators (figure 1 and 2). Total counts ranged between 2 and 3 millions, the acquisition matrix was 128 x 128 with a pixel size of 2.33 mm.

The images were reconstructed by a FBP algorithm and filtered with a Butterworth filter (order = 5; cut off = 0.26). Attenuation correction was performed according to Chang's method (μ = 0.11/cm) [8].

Figure 1. The Alderson phantom filled with ^{123}I-FP-CIT

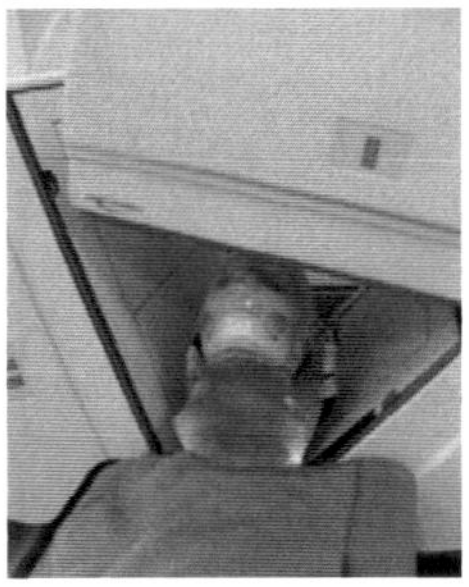

Figure 2. Images acquisition of the Alderson phantom by Philips-Irix.

Reconstructed SPECT data were processed with the "BasGanV2" software, that provides normalized striatal uptake values, and by using a conventional manual ROI method.

The BasGan algorithm performs translations and scale transformation on the bicommissural aligned input image to set the striatal templates, derived from the Talairach atlas, with standard size in an appropriate initial position. An optimization protocol automatically performs fine adjustments in the positioning of blurred templates to best match the radioactive counts, and locates an occipital ROI for background evaluation. Partial volume effect (PVE) correction is also included.

Results are displayed in a report which includes graphical visualization of ratio uptake calculations compared with a reference built-in database (population of 96 subjects with essential tremor from a multicenter study by the AIMN Neurology Study Group).

In the conventional manual method, ROIs were drawn on the head of the caudate, the putamen and the parieto-occipital region choosing the slice where the striatum could be best identified and adding the adjacent upper and lower slices to build a three-slice section. Ratios were calculated, according to

International Conference on Image Optimisation in Nuclear Medicine (OptiNM) IOP Publishing
Journal of Physics: Conference Series **317** (2011) 012004 doi:10.1088/1742-6596/317/1/012004

the same formula which is automatically applied by BasGan (striatum-background/background), based on average counts per voxel. PVE correction was not applied to the manual ROI analysis.

Correlation analysis was performed between true and measured ratios for each method, after checking for normal distribution of the data by using Kolmogorov-Smirnov test.

3. Results

Table 1 presents the true and measured uptake ratios obtained in this study. Good agreement has been found for both methods.

Both methods underestimated almost all the binding values. This could be due to the attenuation correction method based on a mathematic standard algorithm, which maybe is less accurate than the one based on transmission data (i.e. low-energy TC scan). This hypothesis is under investigation. The manual method, however, underestimates values more than BasGanV2, probably due to the absence both of PVE correction and the lack of 3D striata structure information.

Both considered methods have demonstrated good correlation for ^{123}I-FP-CIT SPECT study measurements, as shown in table 2. In fact all the R striata values are high. The lower index value obtained with both methods for the right caudate can be explained by a slightly less accurate filling of this chamber that presents a different and less advantageous morphology with respect to the contralateral, making it difficult to avoid the presence of air bubbles in it.

The BasGanV2 algorithm, however, performed better in determining all considered uptake ratios compared to the manual ROI method evidenced by the higher R index and a better linear trend, as shown in figure 3.

Table 1. True and measured uptake data of phantom studies normalized to the occipital reference ROI.

	Left Caudate			Left Putamen			Right Caudate			Right Putamen		
Experimental session	True values	BasGan values	ROI method values	True values	BasGan values	ROI method values	True values	BasGan values	ROI method values	True values	BasGan values	ROI method values
1	2.7	2.7	1.4	1.0	1.3	0.7	7.8	4.6	2.9	5.0	3.6	2.4
2	3.5	2.7	1.5	3.5	2.3	1.5	5.0	3.4	2.3	5.0	3.0	2.2
3	4.7	4.3	2.6	4.8	4.3	3.0	8.0	6.3	4.2	8.0	6.2	4.6
4	5.7	4.8	3.1	5.7	4.1	2.9	7.8	5.6	3.3	7.8	4.9	3.0
5	6.5	6.6	4.3	6.6	5.7	4.1	8.0	7.0	4.7	8.0	6.5	4.6
6	8.0	7.7	4.9	8.0	7.0	4.6	8.0	7.3	4.4	8.0	6.7	4.7
7	8.7	7.0	5.0	8.7	6.7	4.8	8.7	6.9	4.2	8.7	6.0	3.9
8	8.7	6.2	3.5	4.0	3.5	1.6	8.7	5.8	3.7	4.0	3.1	1.6
9	8.7	6.8	3.7	7.2	5.8	3.3	8.7	6.7	4.3	7.2	5.4	3.7

Table 2. R values obtained in BasGan ($p < 0.001$, except for right C/B where $p < 0.01$) and manual ROI methods ($p < 0.005$, except for right C/B where $p < 0.05$).

Method	Left C/B	Left P/B	Right C/B	Right P/B	Right C/P	Left C/P
BasGanV2	0.93	0.97	0.78	0.93	0.95	0.94
Manual ROI	0.87	0.96	0.72	0.89	0.95	0.93

4. Conclusions

In this study BasGanV2 performed better than the standard ROI method considered, confirming to be an accurate automatic algorithm for 3D semi-quantification of SPECT uptake. The ROI positioning is based on a template derived from Talairach and Tournoux neuroanatomical atlas being user-independent, and, furthermore, the software includes PVE correction and the possibility to utilise a control subjects database.

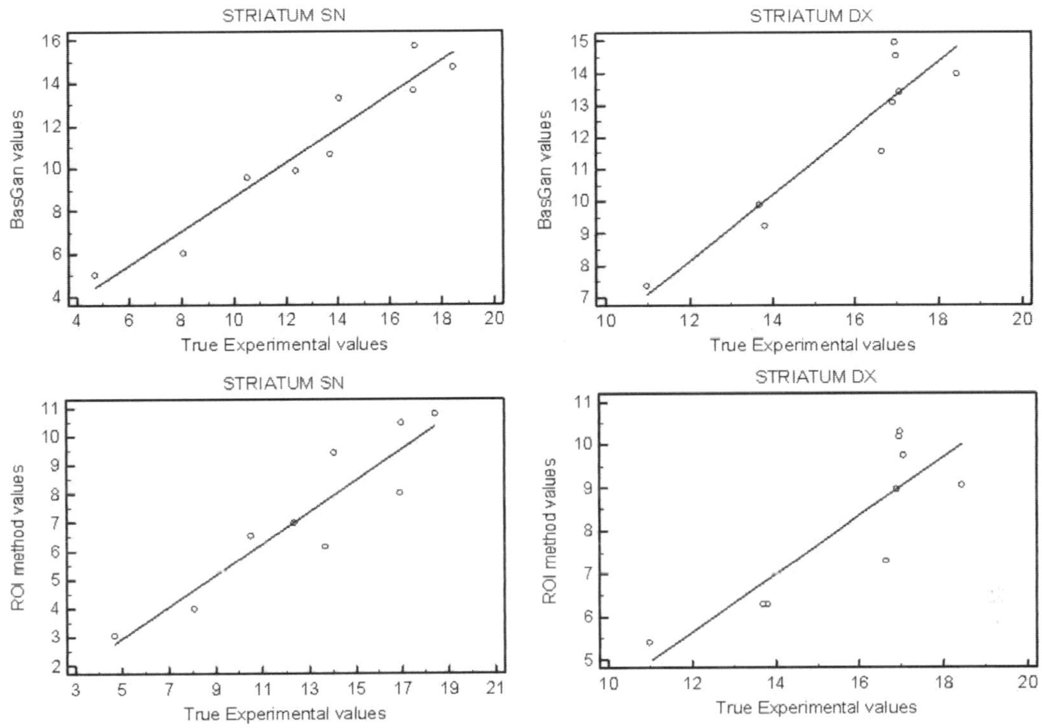

Figure 3. Linear trend of striatum SN and striatum DX for BasGan ($R_{mean} = 0.95$, $p < 0.001$) and ROI method ($R_{mean} = 0.89$, $p < 0.005$), respectively.

Therefore, this new software is suitable for the daily practice, being a simple and accurate method of semiquantitative analysis with the advantage to give reproducible results, simplifying the follow-up of patients and allowing multicenter comparison studies.

References

[1] Booij J, Speelman J D, Horstink M W and Wolters E C 2001 The clinical benefit of imaging striatal dopamine transporters with [123I]FP-CIT SPET in differentiating patients with presynaptic parkinsonism from those with other forms of parkinsonism *Eur. J. Nucl. Med.* **28** 266 –72

[2] Benamer T S *et al* 2000 Accurate differentiation of parkinsonism and essential tremor using visual assessment of [123I]-FP-CIT SPECT imaging: the [123I]-FP-CIT study group. *Mov. Disord.* **15** 503–10

[3] Chouker M, Tatsch K, Linke R, Pogarell O, Hahn K and Schwarz J 2001 *Nucl. Med. Commun.* **22** 721–5

[4] O'Brien J T, Colloby S, Fenwick J, Williams E D, Firbank M, Burn D, Aarsland D and McKeith IG 2004 *Arch. Neurol.* **61** 919–25

[5] Pirker W, Djamshidian S, Asenbaum S, Gerschlager W, Tribl G, Hoffmann M and Brücke T 2002 *Mov. Disord.* **17** 45–53

[6] Italian Association of Nuclear Medicine www.aimn.it

[7] Calvini P, Rodriguez G, Inguglia F, Mignone A, Guerra U P and Nobili F 2007 *Eur. J. Nucl. Med. Mol. Imaging* **34** 1240-53

[8] Darcourt J, Booij J, Tatsch K, Varrone A, Vander Borght T, Kapucu O L, Någren K, Nobili F, Walker Z and Van Laere K 2009 EANM procedure guidelines for brain neurotransmission SPECT using 123-Ilabelled dopamine transporter ligands, version 2 *Eur. J. Nucl. Med. Mol. Imaging* Guidelines

MR-DTI and PET multimodal imaging of dopamine release within subdivisions of basal ganglia

A Tziortzi[1,2], G Searle[2], C Tsoumpas[3], C Long[2], P Shotbolt[2], E Rabiner[2], M Jenkinson[1] and R N Gunn[2]

[1] FMRIB Centre, Department of Clinical Neurology, University of Oxford, UK
[2] GlaxoSmithKline, Clinical Imaging Centre, Imperial College London, UK
[3] Division of Imaging Sciences & Biomedical Engineering, School of Medicine, King's College London, UK

Email: tziortzi@fmrib.ox.ac.uk

Abstract. The basal ganglia is a group of anatomical nuclei, functionally organised into limbic, associative and sensorimotor regions, which plays a central role in dopamine related neurological and psychiatric disorders. In this study, we combine two imaging modalities to enable the measurement of dopamine release in functionally related subdivisions of the basal ganglia. [11C]-(+)-PHNO Positron Emission Tomography (PET) measurements in the living human brain pre- and post-administration of amphetamine allow for the estimation of regional dopamine release. Combined Magnetic Resonance Diffusion Tensor Imaging (MR-DTI) data allows for the definition of functional territories of the basal ganglia from connectivity information. The results suggest that there is a difference in dopamine release among the connectivity derived functional subdivisions. Dopamine release is highest in the limbic area followed by the sensorimotor and then the associative area with this pattern reflected in both striatum and pallidum.

1. Introduction

Dopamine is a neurotransmitter which enables the transmission of signals in the brain. Its abundance or insufficient formation can impair motor, cognitive and emotional functions such as in Schizophrenia and Parkinson's diseases.

In such neurological and psychiatric diseases, unbalanced dopamine formation or release is observed in the basal ganglia, a group of anatomical nuclei whose role has not yet been fully elucidated due to their complex neuropathological mechanisms. Anatomically, basal ganglia consist of: striatum, which is further sub-divided into caudate, putamen and ventral striatum; pallidum which is subdivided into globus pallidus external, globus pallidus internal and ventral pallidum; substantia nigra and the subthalamic nuclei. Basal ganglia are functionally organised into limbic, associative and sensorimotor areas [1]. However, the boundaries of these regions are not defined from anatomical information.

In order to understand the role of dopamine in neurological and psychiatric disorders, several PET studies have measured dopamine release in the brain following the administration of amphetamine. In these studies dopamine release has been measured in anatomical subdivisions of the basal ganglia. Here we propose a multimodal approach and in particular a combined DTI-PET method to measure dopamine release in the functional areas of the basal ganglia (specifically in the striatum and pallidum). Probabilistic tractography methods are applied to obtain connectivity based functional subdivisions of the brain for application to $[^{11}C]$-(+)-PHNO PET data. $[^{11}C]$-(+)-PHNO is an agonist radioligand which binds to the Dopamine D2/D3 receptors and has preferential affinity for the D3 receptors. Due to its agonist profile and its selectivity for the D3 receptors, it has a higher sensitivity to detect dopamine release in comparison to other antagonist radioligands.

2. Methods

2.1. Data acquisition

A total of eleven healthy volunteers were recruited and underwent two $[^{11}C]$-(+)-PHNO PET scans at baseline conditions and after the oral administration of the dopamine releasing agent amphetamine (oral dose 0.3 mg/kg). Furthermore DTI data of 32 directions blip-up and blip-down were acquired for each subject. The study was approved by Essex 1 Research Ethics Committee and the Administration of Radioactive Substances Advisory Committee (ARSAC). All PET and MRI scans were performed at the Clinical Imaging Centre, Hammersmith Hospital, London.

2.2. DTI analysis

DTI data were corrected for motion, eddy-currents and distortions using FSL (FMRIB Software Library, University of Oxford). Subsequently FSL probabilistic tractography, a method that can delineate the neuronal tracts and hence define the probability of connection between brain regions, was applied to derive the connectivity defined functional subdivision (conFROI) of striatum and pallidum. conFROIs were defined using the following projections (a) Limbic: the medial orbital cortex, amygdala and hippocampus are connected to the limbic striatum whose limbic area then projects to the limbic pallidum, (b) Associative: dorsolateral prefrontal cortex projects to the associative striatum which subsequently projects to the associative pallidum (c) Sensorimotor: the primary, premotor and SMA cortices project to the sensorimotor striatum which then projects to the sensorimotor pallidum. The probabilistic tractography took place in each subject's space (diffusion space) and the results were non-linearly registered to the MNI space using the non-linear deformation algorithm FNIRT [2]. The CIC neuroanatomical atlas [3] was used for the definition of the cortical, striatal and pallidal masks.

2.3. PET analysis

PET images were corrected for motion by realigning each frame to a reference frame using mutual information as the cost function. The basis-function implementation of SRTM was applied to the dynamic PET data to derive parametric images of the binding potential (BP_{ND}) which reflects the receptor availability [4]. conFROIs were then applied to the parametric BP_{ND} images to derive the conFROIs BP_{ND} estimates.

2.4. Dopamine release measurements

For each conFROI dopamine release was calculated as the percent change in BP_{ND} elicited by amphetamine (ΔBP_{ND}),

$$\Delta BP_{ND} = \frac{BP_{ND}^{Amph} - BP_{ND}^{Base}}{BP_{ND}^{Base}}\ 100\ \% \qquad (1)$$

where BP_{ND}^{Base} is the BP_{ND} in the baseline condition and BP_{ND}^{Amph} is the BP_{ND} in the amphetamine condition.

3. Results

Across subjects, a topographically consistent subdivision of the basal ganglia was achieved. The in-vivo human connectivity derived functional subdivisions demonstrated a functional organisation similar to that obtained in post-mortem non-human primate tracing studies [1], shown in figure 1. The dopamine release estimates in the conFROIs of striatum and pallidum are presented in figure 2.

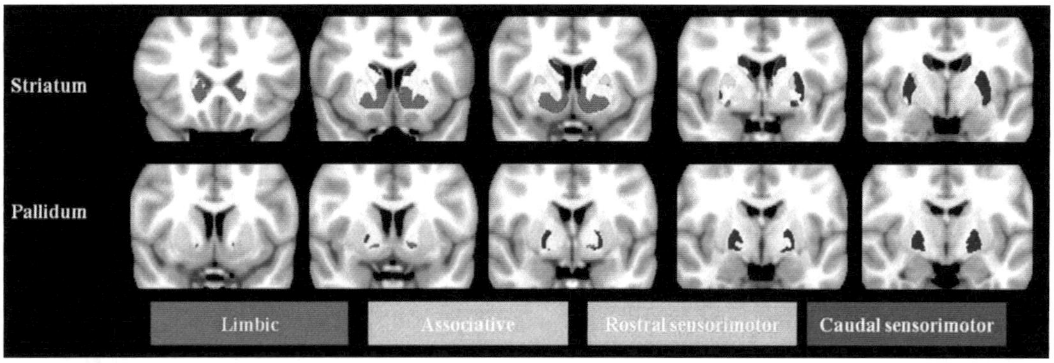

Figure 1. Functional subdivisions (conFROIs) of the BG for a representative subject defined using structural connectivity information from DTI.

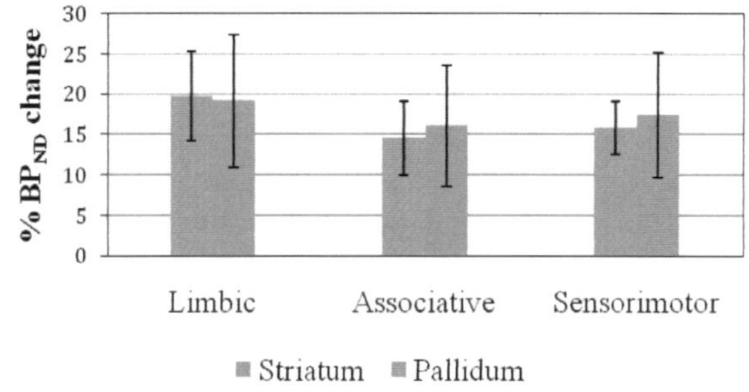

Figure 2. The percentage BP_{ND} change elicited by amphetamine administration.

4. Conclusions

The anatomical connectivity data from DTI were able to produce robust subdivision of the BG that corresponded well to post-mortem data. Dopamine release was highest in the limbic area followed by the sensorimotor and then the associative area with this pattern reflected in both striatum and pallidum.

References

[1] Haber S N 2003 The primate basal ganglia: parallel and integrative networks *Journal of Chemical Neuroanatomy* **26** 317-30

[2] Andersson J, Smith S and Jenkinson M 2008 FNIRT-FMRIB non-linear image registration tool *14th Annual Meeting of the Organization for Human Brain Mapping (HBM 2008: Melbourne, Australia)* Poster: 496

[3] Tziortzi A C, Searle G E, Tzimopoulou S, Salinas C, Beaver J D, Jenkinson M, Laruelle M, Rabiner E A and Gunn R N 2011 Imaging dopamine receptors in humans with [11C]-(+)-PHNO: Dissection of D3 signal and anatomy *NeuroImage* **54** 264-77

[4] Gunn R N, Lammertsma A A, Hume S P and Cunningham V J 1997 Parametric imaging of ligand-receptor binding in pet using a simplified reference region model *Neuroimage* **6** 279-87

International Conference on Image Optimisation in Nuclear Medicine (OptiNM) IOP Publishing
Journal of Physics: Conference Series **317** (2011) 012006 doi:10.1088/1742-6596/317/1/012006

Investigation of the effects of attenuation correction on the compatibility of two single photon emission computed tomography systems with the use of segmentation through registration

M Lampaskis[1,2], K Killilea[1], P Metherall[1], A Harris[1] and D Barber[1]

[1] Department of Medical Physics and Clinical Engineering, University of Sheffield, United Kingdom
[2] Department of Mechanical and Manufacturing Engineering, University of Cyprus, Cyprus

E-mail: lampaskis.marios@ucy.ac.cy

Abstract. The aim of this work was to compare images acquired from two Single Photon Emission Computed Tomography (SPECT), performing attenuation correction using different systems, to evaluate the level at which images from these systems can be used on patients to assess myocardial perfusion. The two systems are the Siemens E-cam with profile attenuation correction and the General Electric Hawkeye system. This study was performed using an anthropomorphic torso phantom. The motivation was to examine if attenuation corrected images from these systems are comparable when assessing the myocardial function of patients, with different conditions regarding background (adjacent tissues) activity and the presence or not of defects on the cardiac wall. To perform the analysis of the acquired images specialized software were used to extract information relating to the activity distribution within the cardiac insert (simulated myocardium). This was based on standardized myocardial segmentation used clinically, by the application of image registration using an artificial reference model. The results show that adjacent tissue activity did not affect the ability to detect defects. Further, the application of attenuation correction may reduce the comparability of the two systems to a small degree.

1. Introduction

The aim of this work was to compare images acquired from two SPECT (which perform attenuation correction using different methods) systems and corrected for the effects of attenuation, to conclude the level at which images from these systems can be used on patients to assess myocardial function. The two systems are the Siemens E-cam with profile attenuation correction located at the Northern General Hospital and the General Electric Hawkeye system located at the Royal Hallamshire Hospital.

2. Materials and methods

2.1. Anthropomorphic torso phantom

The anthropomorphic torso phantom (Bright Technologies Ltd, Sheffield, UK) used to carry out the experiments. It is designed to mimic the properties of a human torso in terms of the attenuation of its various components and their relative positions. Its size and shape are comparable to that of a normal adult male. The main components consist of: two lung inserts filled with Styrofoam, a liver insert which is isolated from other organs and can be filled with different levels of activity, a spine (in the form of a high-density Teflon bar) to simulate the attenuation and scatter that occurs due to the vertebrae, the cardiac insert (only left ventricle) with an array of different defect simulating inserts.

As mentioned above the different chambers of the phantom can be filled with water and suitable levels of activity to simulate the activity distribution that the user wants to test. This is done so as to assess possible scatter as well as attenuation effects/artifacts that may occur in an examination. All fillable chambers used (excluding the lungs) were filled with water. The lung inserts were filled with Styrofoam beads so as to better mimic the attenuation of the lung of a human subject. The cardiac insert was used with and without added defects. The defects used were fillable (full of water) and were placed in the anterior basal region and the inferior mid/apical region of the simulated cardiac wall of the left ventricle.

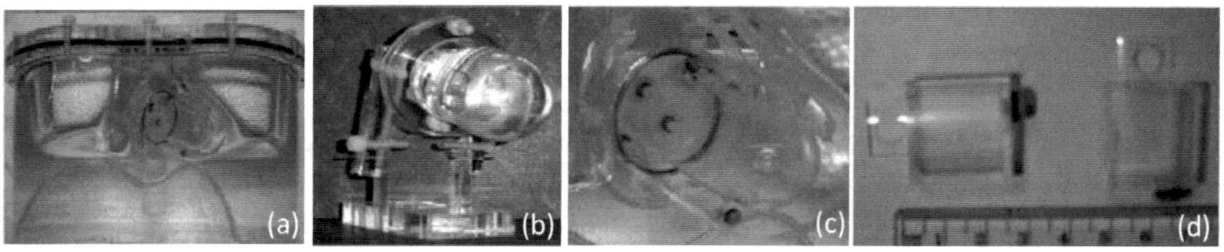

Figure 1. Pictures of (a) the torso phantom from an anterior view, (b) the cardiac insert mounted outside the torso, (c) close up when mounted inside the phantom, (d) the defects used to simulate perfusion abnormalities (scale is in cm).

2.2. General Electric Hawkeye SPECT/Low dose CT and Siemens E-cam profile system

The GE Hawkeye system is situated at the Royal Hallamshire hospital in the Medical Physics and Clinical Engineering Department. It incorporates two gamma camera heads and a low dose CT scanner. This arrangement allows for SPECT images to be acquired without moving the patient, obtaining an attenuation map. Additionally, the image of the scattered radiation is also acquired and can be used for scatter correction. Automatic body contouring (positioning of the camera detectors as close as possible to the imaged volume) is facilitated by moving the two gamma camera heads towards (or away) from the imaged volume to maximize the resolution of the collimator.

The Siemens E-cam system is situated at the Northern General Hospital in the Medical Physics Department. It incorporates two gamma camera heads. To obtain an attenuation map, it uses two arrays of Gadolinium 153 in conjunction with the camera heads. This arrangement allows for the attenuation map and the activity images to be acquired at the same time. This can only be done when the camera heads are in L-mode (i.e. at 90° to each other). Both emission and transmission data can be acquired simultaneously (reducing examination time). The most important limitation is that the attenuation map acquired can only be used for attenuation correction; it is not of a quality that can be used for anatomical information.

2.3. Data acquisition

The phantom was prepared appropriately (as discussed in 2.1) and then the activity (^{99m}Tc in saline) was added to the liver, thorax and cardiac regions. The first acquisition of the session was obtained. Next, additional activity was added to the liver and the last acquisition of the session was performed. Two sessions on each system where held (as shown in table 1), one with defects in the cardiac insert and one without. The phantom was "emptied" before each session.

2.4. Image reconstruction and reorientation

The Ordered Subset Expectation Maximization (OSEM) algorithm [1] algorithm used to reconstruct the acquired data. All data were processed with the same algorithm (OSEM) to avoid differences based on the reconstruction method. In all cases the iterative reconstruction was stopped after three iterations.

The acquired images where reoriented along the long axis of the heart. This was done by an experienced operator, since possible errors at this process could have a large effect in further stages. The images normally are correct to less than 5° deviation from their true position.

Table 1. Acquisition parameters: four sets (A, B, C, D) of data were acquired for each system including the raw emission data and the raw attenuation data.

Acq. par.	Myocardium activity (MBq)	Liver activity (MBq)	Thorax activity (MBq)	Ideal Heart/ Liver ratio	Defects
A	9-11	50-70	30-50	0.17	Anterior & Inferior
B	9-11	180-220	30-50	0.05	Anterior & Inferior
C	9-11	50-70	30-50	0.17	None
D	9-11	180-220	30-50	0.05	None

2.5. Segmentation of the left ventricle

The images were registered to an artificial model that was produced, based on the known geometry of the cardiac insert, to allow extraction of the segment information for the acquired images [2]. A segment number, as shown in figure 2, was assigned to each part of the known-geometry model [3].

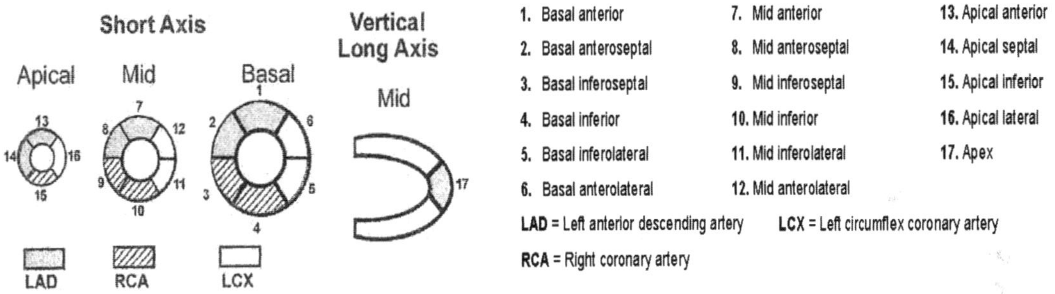

Figure 2. Assignment of 17 myocardial segments to coronary regions.

By segmenting the image, the activity at each part of the cardiac insert could be examined independently. The acquired images were segmented according to different coronary artery regions simulated by the cardiac insert. The average intensity over each segment was obtained and displayed on circumferential polar plots (figure 3), similar to the way commonly performed in clinical practice.

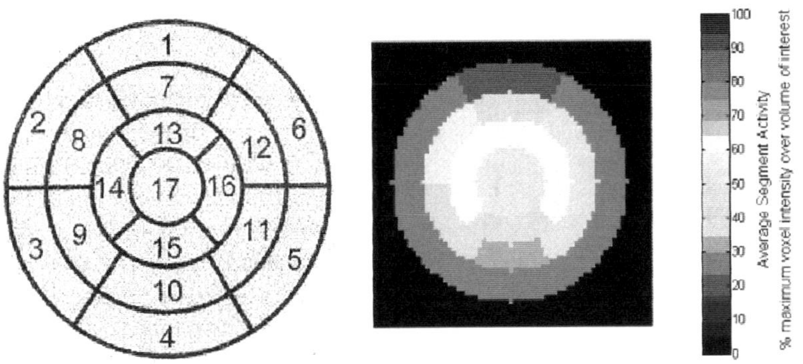

Figure 3. Example of a circumferential polar plot (right) in accordance to the cardiac segmentation map (left), numbered as in figure 2.

3. Results

In total, eight data sets were acquired (four from each system), using the image acquisition parameters shown in table 1. The acquired data where then reconstructed with and without attenuation correction. An example of the attenuation correction effects on a circumferential polar plot can be seen in figure 4. The parameters explored included variation levels of liver activity and the presence of defects in the anterior and inferior regions. By comparing the maximum change that occurred in the average segment intensity, the importance of each factor was evaluated. For example, by quadrupling the liver activity, only 5 - 10% change in any segment was noticed; the ability to detect defect was not affected.

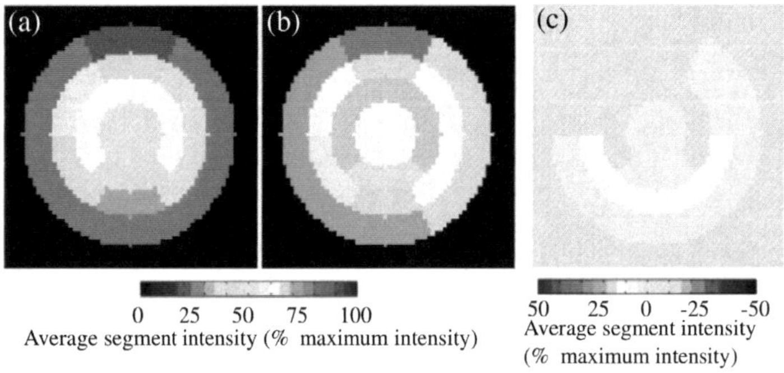

Figure 4. Examples of circumferential polar plots produced (a) without and (b) with attenuation correction. The plot in (c) shows the difference of (a) and (b). Note the increased intensity in the septal region and the presence of defects in the mid anterior and mid inferior.

In figure 5, the effect of attenuation correction on circumferential plots from the two different methods can be observed. Even though the exhibited enhancement of the Siemens E-cam system (figure 5b) is noticeably greater than that of the General Electric Hawkeye system (figure 5a), it cannot be considered significant.

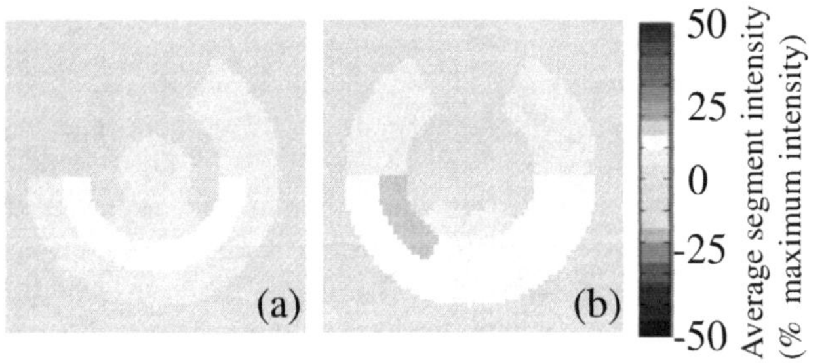

Figure 5. Example: the difference of circumferential polar plots produced with and without attenuation correction for (a) the General Electric Hawkeye SPECT/Low dose CT and (b) the Siemens E-cam Profile system.

4. Conclusion

The results obtained suggest that the application of attenuation correction may reduce the comparability of the two systems only to a small degree. Future work should focus on acquiring a largerer amount of data to explore the possibility of using a dynamic phantom where the myocardium and diaphragm motion is simulated.

References

[1] Bateman T M and Cullom S J 2005 *Seminars in Nuclear Medicine* **31** 37-51
[2] Barber D C and Hose D R 2005 *Journal of Medical Engineering & Technology* **29** 53-63
[3] Cerqueira M D *et al* 2002 *Circulation* **105** 539-542

Diagnostic improvement based on image processing in low extremities inflamation scintigraphy

M Lyra[1,2], S Kordolaimi[1] and A L Salvara[1]

[1] Radiation Physics Unit, A΄ Radiology Department, University of Athens, Aretaieion Hospital, GR-11528, Athens, Greece
[2] Division of Nuclear Medicine, A' Radiology Department, University of Athens, Aretaieion Hospital, GR-11528, Athens, Greece

E-mail: mlyra@medimaging.gr

Abstract. The purpose of this study is the improvement in evaluation of the inflammation extent on the scintigraphic imaging, by utilizing statistical indices (Inflammation Projection Ratio - IPR, skewness, kurtosis and Mean Pixel Value - MPV). Image analysis was performed by means of an Interactive Data Language (IDL) tool. Twelve patients were referred for a radionuclide (Tc99m- Leukoscan) scan, by a GE Healthcare gamma camera, on the suspicion of an infectious lesion in the extremities. The findings of the study are that pathological tissues have a higher IPR index (3.12 to 4.32) compared to normal tissue (~ 1). Furthermore, MPV, skewness and kurtosis differ significantly (> 5%) from normal to inflammable extremities. As a conclusion, image processing provides effective information in the structure and facilitates diagnosis semi-quantitatively.

1. Introduction

Imaging of inflammatory processes and infection is a form of tissue characterization by nuclear medicine [1, 2]. This derives from the fact that scintigraphic detection of infection and inflammation is based on functional tissue changes. Thus, it is preferred from other radiological techniques which merely rely on anatomical changes [3-5].

The agent utilized in this study is Tc99m-labeled Fab' fragment of IMMU-MN3, which is an immunoglobulin GI murine monoclonal antibody. It can detect infection/inflammation with an accuracy that is comparable to white blood cell (WBC) scanning [6-10].

In this study, an improvement in the evaluation of the inflammation extent on the scintigraphic imaging was attempted, by determination of the index Inflammation Projection Ratio (IPR) and the intensity isocontours creation on the images. Furthermore, statistical analysis was attempted by utilizing mean pixel value (MPV), skewness, and kurtosis to create indicative indices.

2. Materials and methods

2.1. Monoclonal antibody preparation and labeling

Tc99m-Sulesomab, that is Tc99m- labeled Fab' fragment of IMMU-MN3, was used to obtain images of both lower extremities of each patient, produce indices from the measured counts on the acquired images and make cartographies of the disease local extension by intensity isocontouring of the normal and inflamed extremities.

The Fab' fragment is provided in a ready to label lyophilized kit from Immunomedics, (Leukoscan) [11]. Labelling was accomplished by adding approximately 1000 to 1500 MBq of Tc99m-pertechnetate in saline directly into the vial containing 1.25 mg of the monoclonal antibody Fab' fragment. The injected material contains 0.25 mg protein labelled with 555 - 925

MBq (15 to 25 mCi) of Tc99m. It is prepared using a 5 minute, 1 step procedure with a labelling efficiency greater than 99%. Tc99m- LeukoScan has a biological half-life of 10.5 hours. The labelled antibody is injected intravenously over 30 seconds. Tc99m- Anti-Granulocyte Fab' normally concentrates in the liver, spleen, kidneys, and bladder.

2.2. Acquiring images
Twenty-minute whole body planar images of high statistics, at 3 hours, 8 hours and 24 hours post injection of radiopharmaceutical Tc99m- Sulesomab were acquired by a GE Healthcare gamma camera connected to a GE healthcare dedicated processor Xeleris-2. Total counts of each planar extremeties' image were more than 400 kcounts. The distribution is evaluated using the information from statistics retrieved from anterior and posterior scintigrams.

2.3. Inflammation projection ratio
Three regions of interest (ROI) were drawn on each one of the spot images of the lower extremities: one that encompassed the whole pathologic area, a ROI at a corresponding area of the normal extremity and a ROI for background counts for the evaluation of the projection of the infectious area. Image analysis by normalized ROIs is used to determine pathologic area uptakes (see figure 1). From the measured counts, an index, Inflammation Projection Ratio (IPR), representing the projection of the lesion for each patient was calculated. The IPR indices, for all 12 patients, were calculated for the 3 series of data (3h, 8h, 24h). The indices represent the ratio of the counts in the pathologic area to the counts of a normal area similar to the infected tissue, after subtraction of normalized background counts, obtained from a normal soft tissue area.

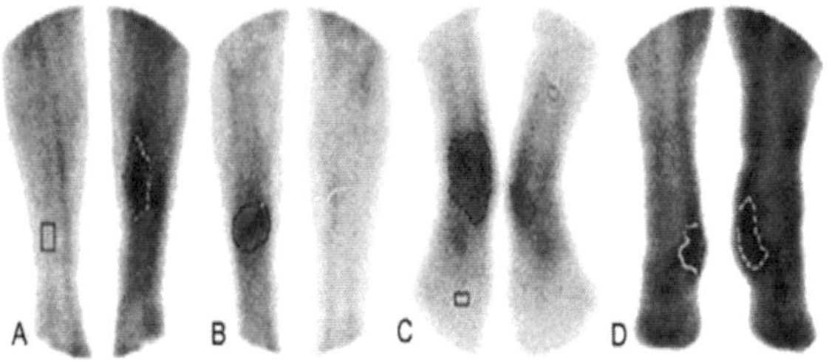

Figure 1. Typical regions of interest (ROIs) of pathological and normal extremities and a background ROI for the lower extremities (A) in the left tibia, (B) in the right tibia, (C) in the right knee, and (D) in the left foot.

2.4. Inflammation intensity isocontours- image processing
Image analysis is performed by means of an Interactive Data Language (IDL) tool, which is a platform independent data analysis tool with broad range of mathematical and graphical capabilities.

Local extent of the inflammation is obtained by the intensity isocontours on the inflammation region created by IDL programming. A region of interest (ROI) is an area of an image defined for further analysis or processing. CONTOUR is a routine that enable the programmed definition of ROIs. It traces the outlines of thresholded ROIs (see figure 2) [12, 13].

Image Statistics calculated with the IDL mathematical operators are the sum, mean, standard deviation, variance of the pixel values, skewness and kurtosis of an image. Skewness, kurtosis and mean pixel value were found to be useful characteristics of the images that show not only qualitatively but in a semi-quantitative measure the extent of inflammation localization.

Skewness is defined as a measure of symmetry, or more precisely, the lack of symmetry. Skewness is zero if the data are distributed symmetrically around the mean, negative if the data are more spread out on the left of the mean, and positive if the data are more spread out on the right of the mean [14, 15]. Kurtosis is a measure of whether the data are peaked or flat relative to a normal distribution [14].

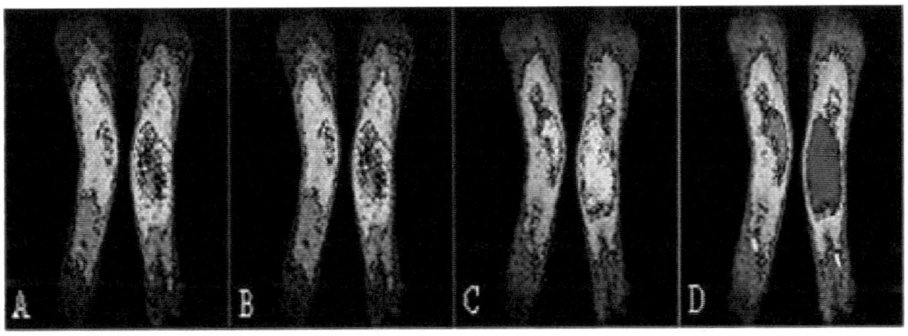

Figure 2. ROIs traced by isocontour analysis with IDL programming. The local extent of the inflammation is obtained by the intensity isocontours on the inflammation region creating by IDL programming.

3. Results

3.1. IPR estimates

The three sets of Inflammation Projection Ratios (IPR) calculated for each patient are shown in table 1. The IPRs values, obtained by the 24h post injection Tc99m- Leukoscan images, seem to be the higher values that give a greater range for inflammation evaluation. By analyzing the data, it is shown that pathological tissues have a higher IPR index compared to normal tissues which is clearly depicted in figure 3. More specifically, the pathological IPR ranges from 3.12 to 4.32 while the IPR of the healthy tissue should be approximately 1 (or a little more owing to the inhomogeneities of the human body).

Table 1. Inflammation projection ratio (IPR) patient specific/inflamemation focus.

Patient number	IPR_{3h}	IPR_{8h}	IPR_{24h}	$IPR_{aver} \pm SD_{IPR}$	$IPR_{totaver}$
1	4.1	4.2	4.8	4.37 ± 0.22	3.72 ± 0.60
2	3.8	4.3	4.7	4.27 ± 0.26	
3	3.5	4.2	4.5	4.07 ± 0.30	
4	4.9	4.7	5.1	4.90 ± 0.12	
5	4.2	3.9	4.4	4.17 ± 0.15	
6	3.3	3.4	3.8	3.50 ± 0.15	
7	3.1	3.1	3.4	3.20 ± 0.10	
8	3.5	3.3	3.7	3.50 ± 0.12	
9	2.8	2.6	3.1	2.83 ± 0.15	
10	3.3	3.1	3.9	3.43 ± 0.24	
11	2.9	2.7	3.3	2.97 ± 0.18	
12	3.3	3.5	3.4	3.40 ± 0.06	
Normal patient 1	1.17	1.15	1.15	1.16 ± 0.007	1.11 ± 0.04
Normal patient 2	1.15	1.05	1.013	1.07 ± 0.04	

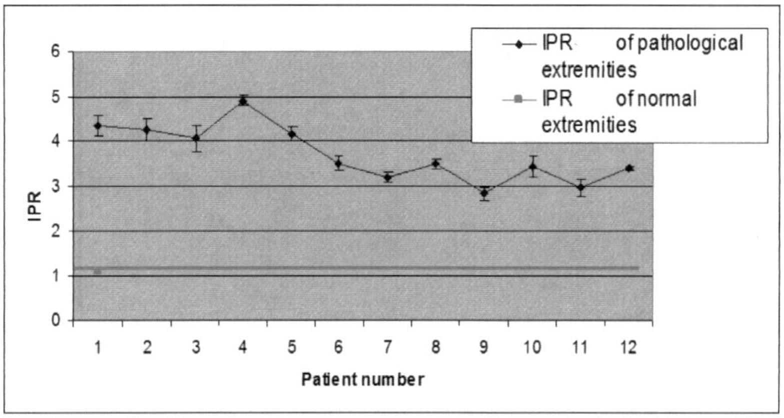

Figure 3. Graphic display of IPR index for normal and pathological tissues after Tc99m- leukoscan injection.

3.2. Isocontours analysis

The isocontouring and statistical analysis of Tc99m- Leukoscan images accentuates the extension of inflammation region. The comparison of skewness and kurtosis of abnormal and normal part of the image gives a numerical grade of the data population of the inflammation presented by Tc99m- Leukoscan imaging (see table 2). By calculating the t-value from 'Student's' t-test (for paired samples), it is estimated that skewness and kurtosis both equal to 12, while the theoretical value for t, defining the level of significance in 5%, equals to 2.23. The calculated value outnumbers the theoretical value which means that the null hypothesis (the two populations are equal) is rejected. So, skewness and kurtosis differ from pathological to healthy tissues significantly.

Apart from skewness and kurtosis, another index evaluated is the mean pixel value (MPV). Table 3 compares the MPV of a ROI in the pathological region to the corresponding in the normal region. By calculating the p- value, it is clearly shown that this index changes significantly in case of infection.

Table 2. Statistical analysis of Tc99m- leukoscan images – Skewness/Kurtosis.

Patient number	Skewness Normal extrem.	Skewness Inflam. Extrem.	Kurtosis-3 Normal extrem.	Kurtosis-3 Inflam. Extrem.
1	-1.22	-0.84	1.62	0.05
2	0.39	0.37	-1.14	-1.20
3	-1.59	-1.65	2.03	2.16
4	-0.63	-0.28	-0.98	-1.60
5	-0.59	-0.65	-1.06	-1.14
6	-1.04	-0.70	-0.20	-1.19
7	-0.50	-0.38	-1.06	-1.43
8	-0.59	-0.65	-1.06	-1.14
9	-0.84	-0.48	-0.57	-1.43
10	-0.94	-0.87	-0.53	-0.75
11	-1.19	-1.19	-0.05	0.32
12	-0.78	-0.76	-0.30	-0.40

Table 3. Statistical analysis of Tc99m- leukoscan images – Mean pixel value.

Patient number	Normal extremities Mean Value	Normal extremities SD	Inflammable Extremities Mean Value	Inflammable Extremities SD	p- value
1	192	62	183	70	<0.0001
2	73	70	77	74	0.0249
3	217	54	217	55	0.0002
4	162	91	147	101	<0.0001
5	190	61	173	81	<0.0001
6	211	54	179	90	<0.0001
7	42972	19445	40376	21786	0.001
8	190	61	173	81	<0.0001
9	204	56	167	92	<0.0001
10	198	68	191	75	0.0006
11	202	63	195	66	0.0011
12	200	58	199	59	0.0024

4. Discussion

As far as the projection of the infectious lesion is concerned, the calculated IPR shows that Tc99m- Leukoscan presents high lesion to normal area ratio and more specific comparing to other radiopharmaceuticals [16]. This index, IPR, could also be used for quantitative analysis and for the differentiation of healthy and infected tissue or for the extraction of various functional parameters. The calculated indices IPR in combination with the processed isocontouring images, contribute to the conclusion that Tc99m- Sulesomab scintigrams are characterized by high projection ratios specific for each case and demonstrate the full extent of the inflamed area in the lower extremities, with great accuracy. IPR seems to be a reliable index to determine whether a tissue is pathological or not, although the study of more patients is necessary in order to confirm this conclusion.

Furthermore, skewness, kurtosis and mean pixel value are reliable indices in order to give a numerical grade of the data population of the inflammation presented by Tc99m- Leukoscan imaging. Skewness and kurtosis indices change radically with a slight variation in the sample population especially when the sample is small and care should be taken when using these indices.

References

[1] James S L and Davies A M 2006 *Eur. J. Radiol.* **58** 27–40
[2] Prandini N, Lazzeri E, Rossi B, Erba P, Parisella M G, and Signore A 2006 *Nucl. Med. Commun.* **27** 633–44
[3] Auler M A, Bagg S and Gordon L 2007 *Semin. Roentgenol.* **42** 117–21
[4] Becker W and Meller J 2001 *The Lancet Infectious diseases* **1** 326-33
[5] Love C and Palestro C J 2004 *J. Nucl. Med. Tech.* **32** 47–57
[6] Becker W, Goldenberg D M and Wolf F 1994 *Semin. Nucl. Med.* **24** 1–13
[7] Becker W *et al* 1996 *Clin. Orth. Rel. Res.* **329** 263–72
[8] Pakos E E, Koumoulis H D, Fotopoulos A D and Ioannidis J P A 2007 *Radiol.* **245** 732-41
[9] Pakos E E, Trikalinos T A, Fotopoulos A D and Ioannidis J P A 2007 *Radiol.* **242** 101-8
[10] Vicente A G *et al* 2004 *Clin. Nucl. Med.* **29** 781-5
[11] Australian Nuclear Science and Technology Organization 2002 *LeukoScan Kit: Preparation of Technetium-99m Labelled LeukoScan* (New Jersey: Immunomedics Inc.)
[12] IDL Tutorial 2008 *ITT Visual Information Solutions: Advanced Image Processing* http://www.ittvis.com/
[13] IDL 2009 *ITT Visual Solutions: Image Processing* v. 7.1
[14] NIST/SEMATECH *e-Handbook of Statistical Methods* http://www.itl.nist.gov/div898/handbook/eda/section3/eda35b.htm
[15] Motoyoshi I, Nishida S, Sharan L and Adelson E H 2007 *Nature* **447** 206-9
[16] Lyra M, Frantzis A and Limouris G S 2001 *Radiopharmaceuticals for inflammation /infection* eds G S Limouris, Biersack H J and Shukla S K (MEDITERRA-Pub.) pp 118-28 http://www.medimaging.gr/tutorials/RADIATION%20DOSIMETRY%20 IN %20INFECTION%20Lyra%20et%20al%202001.pdf

Current activities in the ICRP concerning estimation of radiation doses to patients from radiopharmaceuticals for diagnostic use

S Mattsson[1], L Johansson[2], S Leide-Svegborn[1], J Liniecki[3], D Nosske[4], K Riklund[5], M Stabin[6] and D Taylor[7]

[1] Medical Radiation Physics, Department of Clinical Sciences Malmö, Lund University, Skåne University Hospital Malmö, SE-205 02 Malmö, Sweden
[2] Radiation Physics, Department of Radiation Sciences, Umeå University, SE-901 85 Umeå, Sweden
[3] Department of Nuclear Medicine, Medical University, PL-92216 Lodz, Poland
[4] Bundesamt für Strahlenschutz, Fachbereich Strahlenschutz und Gesundheit, Ingolstädter Landstr. 1, DE-85764 Oberschleißheim, Germany
[5] Diagnostic Radiology, Department of Radiation Sciences, Umeå University, SE-901 85 Umeå, Sweden
[6] Department of Radiology and Radiological Sciences, Vanderbilt University, Nashville, TN 37232-2675, USA
[7] School of Chemistry, Cardiff University, Cardiff, CF10 3TB Wales, UK

E-mail: soren.mattsson@med.lu.se

Abstract. A Task Group within the ICRP Committees 2 and 3 is continuously working to improve absorbed dose estimates to patients investigated with radiopharmaceuticals. The work deals with reviews of the literature, initiation of new or complementary studies of the biokinetics of a compound and dose estimates. Absorbed dose calculations for organs and tissues have up to now been carried out using the MIRD formalism. There is still a lack of necessary biokinetic data from measurements in humans. More time series obtained by nuclear medicine imaging techniques such as whole-body planar gamma-camera imaging, SPECT or PET are highly desirable for this purpose. In 2008, a new addendum to ICRP Publication 53 was published under the name of ICRP Publication 106 containing biokinetic data and absorbed dose information to organs and tissues of patients of various ages for radiopharmaceuticals in common use. That report also covers a number of generic models and realistic maximum models covering other large groups of substances (e.g. "^{123}I–brain receptor substances"). Together with ICRP Publication 80, most radiopharmaceuticals in clinical use at the time of publication were covered except the radioiodine labeled compounds for which the ICRP dose estimates are still found in Publication 53. There is an increasing use of new radiopharmaceuticals, especially PET-tracers and the TG has recently finished its work with biokinetic and dosimetric data for ^{18}F-FET, ^{18}F-FLT and ^{18}F-choline. The work continues now with new data for ^{11}C-raclopride, ^{11}C-PiB and ^{123}I-ioflupan as well as re-evaluation of published data for ^{82}Rb-chloride, ^{18}F-fluoride and radioiodide. This paper summarises published ICRP-information on dose to patients from radiopharmaceuticals and gives some preliminary data for substances under review.

1. Introduction

As early as in 1968, Professor R. E. Ellis was asked by the ICRP to prepare a report on the protection of the patient in radionuclide investigations. The report describes the basic principles for minimising the dose to patients receiving radiopharmaceuticals, and also presents a compilation of the absorbed doses resulting from the administration of the more commonly used compounds in nuclear medicine at that time (ICRP Publication 17, [1]). In view of the growing importance of nuclear medicine and the need for more detailed absorbed dose estimates, a group was appointed early in the 1980s as a Task Group of ICRP Committee 2 and later converted to a Joint Task Group with Committee 3 to:

- provide biokinetic models for current and new radiopharmaceuticals which have come into use since Publication 17,
- supply estimated absorbed doses and effective dose equivalents (later effective doses) to patients, including the range of variation to be expected in pathological states, for adults, children, the embryo and foetus.

As a result of the work of the Task Group, ICRP published in 1987 a report entitled 'Radiation dose to patients from radiopharmaceuticals' (Publication 53) [2]. This report contained calculations of absorbed doses and effective dose equivalents per unit activity administered for some 120 radiopharmaceuticals in regular diagnostic use at the time. A first addendum to Publication 53 was included in ICRP Publication 62 [3], where also an age related bladder voiding model was introduced. Publication 62 contained biokinetic and dosimetric (absorbed doses and effective doses) data for six new radiopharmaceuticals, and a table of effective doses per unit administered activity for those radiopharmaceuticals that had been covered in Publication 53. In a second addendum to Publication 53 (ICRP Publication 80) [4], the Task Group presented biokinetic and dosimetric data on 10 new radiopharmaceuticals, and recalculations of dose data for 19 of the most frequently used radiopharmaceuticals in Publication 53 as well as an updated effective dose table. In 2008, a third printed addendum to ICRP Publication 53 was published under the name of ICRP Publication 106 [5] containing biokinetic data and absorbed dose information to organs and tissues of patients of various ages for 25 individual radiopharmaceuticals or groups of radiopharmaceuticals in common use. That report also covers a number of generic models and realistic maximum models covering large groups of substances (e.g. "^{123}I–brain receptor substances"). Together with ICRP Publication 80, most radiopharmaceuticals in clinical use at the time of publication were covered except the radioiodine labeled compounds for which the ICRP dose estimates are still those found in Publication 53. ICRP Publication 106 also includes recommendations related to breast-feeding for mothers who have undergone nuclear medicine investigations.

There is an increasing use of new radiopharmaceuticals, especially substances for PET- and SPECT-investigations and the TG has recently finished its work with biokinetic and dosimetric data for O-(2-[^{18}F]fluorethyl)-L-tyrosine (^{18}F-FET), 3'-deoxy-[^{18}F]-3'-flourothymidine (^{18}F-FLT) and ^{18}F-choline. The work continues now with new data for ^{11}C-raclopride, ^{11}C-PiB and ^{123}I-ioflupan as well as re-evaluation of published data for ^{82}Rb-chloride, ^{18}F-fluoride and radioiodide. The aim of this paper is to describe the current and planned work within the Task Group with the intention to encourage feed-back and input from the nuclear medicine, medical physics and radiation protection communities.

2. Materials and methods

In the reports [2-5], data on each substance are presented in three subsections: biokinetic model, biokinetic data, and table of absorbed dose per unit of activity administered. Unless otherwise stated, the model refers to intravenous administration.

2.1. Biokinetic and dosimetric data

The calculations are based on biokinetic models and best estimates of biokinetic data for individual radiopharmaceuticals. Furthermore for consistency, standardised biokinetic or dosimetric models are used for the dose calculations. For the bladder, a voiding period of 3.5 hours is used. For the gastrointestinal (GI) tract, the former ICRP G-I tract model [6] has hitherto been applied. A kinetic model for substances excreted via the liver and gallbladder has been introduced and used. For activity distributed in circulating blood, the fractional blood volume, or, for short-lived radionuclides, the fractional cardiac output is considered for assessing the distribution of the radioactive source in the body [7].

The rates of the biological processes, for example uptake, metabolism, and excretion, are usually given as the half-time of the corresponding exponential function. If the process is assumed to be multi-exponential, the fraction (*a*) of the organ content belonging to each exponential component is given in brackets immediately after the half-time figure. When rates were reported in the cited publications as fractions per time unit (k), they are transformed into half-times; T = 0.693/k.

The tables sometimes contain empty spaces under the headings T and *a*, usually because the kinetics are described by a complex exponential, or non-exponential, expressions, which cannot easily be defined in the table. This is the case, for example, for activity in the gastrointestinal tract, the gallbladder, and the urinary bladder. In these cases, the tables present only the cumulated activities together with the fractional distributions.

The relative cumulated activities are presented in hours (h). Average organ or tissue absorbed doses are given as milligray (mGy) per megabecquerel (MBq). The effective dose is given as millisievert (mSv) per MBq.

Dose calculations have been performed for adults and 15-, 10-, 5-, and 1-year-old children. The organs (or tissues) are presented in alphabetical order except ‘Remaining organs’, which is placed at the end. The dose to organs or tissues not mentioned in the table can usually be approximated with the value given for ‘Remaining organs’.

For the dose calculations, the MIRD method has been used based on the mathematical phantom and hence S-values from MIRD Pamphlet 11 [8]. For children, absorbed fractions from Cristy and Eckerman [9] were utilised for the dose calculations. For bone tissue dosimetry, a refined model from Eckerman and Stabin [10] was used.

2.2. Effective dose from various radiopharmaceuticals

The quantity ‘effective dose’ [11, 12] can be of practical value for comparing the relative doses related to stochastic effects from: different diagnostic examinations and interventional procedures; the use of similar technologies and procedures in different hospitals and countries; and the use of different technologies for the same medical examination provided that the representative patients or patient populations for which the effective doses are derived are similar with regard to age and gender. However, comparisons of effective doses are inappropriate when there are significant dissimilarities between the age and gender distributions of the representative patients or patient populations being compared (e.g. children, all females, elderly populations) and the Commission’s reference distribution of both genders and all ages. This is a consequence of the fact that the magnitudes of risk for stochastic effects are dependent on age and gender. Risk assessment for medical uses of ionising radiation is best evaluated using appropriate risk values for the individual tissues at risk, and for the age and gender distribution of the population groups undergoing the medical procedures.

Up to now, the Task Group has calculated effective doses according to ICRP Publication 60 [12]. For the exposure of young children, the risk would be higher, perhaps by a factor of two or three [12]. For many common types of diagnostic examinations, the higher risk will be offset by the reduction in administered activity relative to that to an adult. For an age at exposure of approximately 60 years, the risk would be lower, perhaps by a factor of three and will decrease somewhat thereafter [12]. As indicated above, the specific demographics of the medically exposed population raise issues related to the application of the concept of effective dose as a tool for comparing doses from medical irradiation with other sources of exposure to humans. Nonetheless, the quantity continues to be calculated for diagnostic pharmaceuticals and its use for summarizing and comparing dose between radiopharmaceuticals and other procedures involving ionising radiation has been widely accepted in the clinical community.

Journal of Physics: Conference Series **317** (2011) 012008 doi:10.1088/1742-6596/317/1/012008

3. Results and discussion

Table 1 gives examples of effective doses per unit administered activity (E/A_0) for adults for some commonly used radiopharmaceuticals [2, 4, 5]. To give the reader an idea about the effective dose per investigation, these values have been multiplied with "typical" activities (A_0) used for investigations of adults.

For substances labelled with radionuclides of short physical half-life, the variation of E/A_0 is limited after an i.v. injection. For ^{11}C-substances, E/A_0 varies between 0.0027 and 0.0084 mSv/MBq (around a factor of 3). A "realistic maximum model" [5] for ^{11}C, which could be used for those ^{11}C-substances for which there are not yet better estimates, is defined and gives 0.011 mSv/MBq, or about 1.3 to 3 times the effective doses predicted by the compound-specific ^{11}C models.

For ^{18}F-labelled substances, E/A_0 varies between 0.016 and 0.028 mSv/MBq (less than a factor of 2), where the higher value itself is a conservative estimate for a group of brain receptor substances.

Also for ^{99m}Tc-labelled substances, the range of E/A_0-values is limited. Values between 0.0047 and 0.017 mSv/MBq are found.

Table 1. Effective dose per unit administered activity (E/A_0) for adults and the "typical" activities (A_0) used for adults investigations with the corresponding effective dose (E), for some commonly used radiopharmaceuticals.

Radiopharmaceutical	E/A_0 (mSv/MBq)	ICRP Publ.	A_0 [a] (MBq)	E (mSv)
^{3}H-neutral fat, free fatty acids	0.22	80	0.1	0.022
^{11}C-acetate	0.0035	106	500-1 000	1.8-3.5
^{11}C-amino acids (generic model)	0.0056	106	200-400	1.1-2.2
^{11}C-choline	0.0047	TG [b]	400	1.9
^{11}C-brain receptor substances (generic model)	0.0043	106	400	1.7
^{11}C-methionine	0.0084	106	200-400	1.7-3.4
^{11}C-thymidine [methyl-^{11}C]thymidine	0.0035	80	400	1.4
^{11}C-thymidine [2-^{11}C]thymidine	0.0027	80	400	1.1
^{11}C (realistic maximum model)	0.011	106	400	4.4
^{14}C-neutral fat, free fatty acids	2.1	80	0.1	0.21
^{14}C-urea (normal/*Helicobacter* pos.)	0.031/0.081	80	0.1-0.2	0.003-0.016
^{15}O-water	0.0011	106	400	0.44
^{18}F-amino acids (generic model)	0.023	106	400	9.2
^{18}F-brain receptor substances (generic model)	0.028	106	400	11
^{18}F-choline	0.019	TG	400	7.6
^{18}F-FDG	0.019	106	250-400	4.8-7.6
^{18}F-FET	0.017	TG	400	6.8
^{18}F-fluoride	0.024	53/80	400	9.6
^{18}F-L-dopa	0.025	106	400	10
^{18}F-FLT	0.016	TG	400	6.4
^{51}Cr-EDTA	0.0020	80	3-4	0.006-0.008
^{67}Ga-citrate	0.10	80	150	15
^{75}Se-HCAT	0.69	80	0.02; 0.4 [c]	0.01; 0.28 [c]
^{82}Rb-chloride	0.0017 [d]	TG	2000	3.4
^{99m}Tc-apticide	0.0047	106	750	3.5
^{99m}Tc-colloids (large)	0.0094	80	80-500	0.75-4.7
^{99m}Tc-colloids (small), intratum. inj.	0.0012	106	20-100	0.024-0.12
^{99m}Tc-DMSA	0.0088	80	100	0.88
^{99m}Tc-DTPA	0.0049	80	300	1.5

^{99m}Tc-EC	0.0063	106	100-1000	0.63-6.3
^{99m}Tc-ECD	0.0077	106	100-1000	0.77-7.7
^{99m}Tc-furifosmin (rest/exercise)	0.010/0.0089	106	300-900	2.7-9.0
^{99m}Tc-HM-PAO	0.0093	80	500-1000	4.7-9.3
^{99m}Tc-IDA derivatives	0.017	80	150	2.6
^{99m}Tc-MAA	0.011	80	100-200	1.1-2.2
^{99m}Tc-MAG3	0.0070	80	70-200	0.49-1.4
^{99m}Tc-markers, non-absorbable (fluids/solids) *per os*	0.019/0.024	80	10-40	0.19-0.96
^{99m}Tc-MIBI (rest/exercise)	0.0090/0.0079	80	300-900	2.4-8.1
^{99m}Tc-monoclonal antibodies: Intact ab/F(ab′)$_2$-fragm/F(ab′)-fragm	0.0098/0.0097/ 0.011	106	750	7.3-8.3
^{99m}Tc-pertechnetate, without blocking	0.013	80	100-800	1.3-10
^{99m}Tc-pertechnetate, with blocking	0.0042	80	100-800	0.42-3.4
^{99m}Tc-phosphates and phosphonates	0.0057	80	600	3.4
^{99m}Tc-RBC	0.0070	80	800	5.6
^{99m}Tc-pertechnegas/-Technegas	0.012-0.015	80	30	0.36-0.45
^{99m}Tc-tetrofosmin (rest/exercise)	0.0080/0.0069	106	600	4.1-4.8
^{99m}Tc-WBC	0.011	80	200	2.2
^{111}In-monoclonal antibodies: Intact ab/F(ab′)$_2$-fragm/F(ab′)-fragm	0.22/0.20/0.20	106	200	40-44
^{111}In-octreotide	0.054	106	170	9.2
^{123}I-iodide, 35% thyroid uptake	0.22	53/80	2-20	0.44-4.4
^{123}I-iodide, after ablation, 1% thyroid uptake	0.02	53	200-400	4-8
^{123}I-fatty acids (BMIPP/IPPA)	0.016	106	150	2.4
^{123}I-brain receptor substances (generic model)	0.050	106	180	9.0
^{123}I-MIBG	0.013	80	400	5.2
^{123}I-monoclonal antibodies: Intact ab/F(ab′)$_2$-fragm/F(ab′)-fragm	0.026/0.019/0.017	106	200	3.4-5.2
^{131}I-iodide, 35% thyroid uptake	24	53/80	0.2	4.8
^{131}I-iodide, after ablation, 1% thyroid uptake	0.5	53	40	20
^{131}I-iodo hippurate	0.052	80	0.2	0.010
^{131}I-monoclonal antibodies: Intact ab/F(ab′)$_2$-fragm/F(ab′)-fragm	0.42/0.14/0.11	106	200	22-84
^{201}Tl-chloride	0.14	106	80	11

[a] Activities currently used in some places and they should not be interpreted as a recommendation of the optimal activity to be administered. The ranges of "typical" activities used are in some cases explained by different indications for the investigations.

[b] TG means preliminary estimate by the Task Group.

[c] Measured with whole-body counter and gamma camera, respectively.

[d] Interim value based on [14] and [15].

For radiopharmaceuticals where the radionuclide has a longer physical half-life, the differences between various substances are larger and more dependent on the biokinetic behaviour of the substances.

For ^{82}Rb-chloride, now frequently used for myocardial perfusion imaging using PET/CT-technology, a re-evaluation of the "worst-case" dosimetry given in Publication 53 and repeated in Publication 80 is under way. ICRP Publication 53 assumes a model based on relative blood flow to various tissues as a proportion of cardiac output. More recent work within the Task Group has been based on the blood volume and blood flow models developed by Leggett et al. [14] and some direct measurements in humans [15].

For ^{123,124,125,131}I-iodide there is an urgent need to re-evaluate existing biokinetic and dosimetric models. This will be done in parallel to the ICRP-work related to occupational exposure by the different radioiodine isotopes. As soon as new S-values for the new ICRP reference voxel phantoms for adults [13] and for children of various ages are available, the Task Group will use them for improved organ/tissue dose estimates. The new human alimentary tract (HAT) model [16] will also be used. Effective doses now calculated according to ICRP Publication 60 will then also be given according to ICRP Publication 103 [11].

Preliminary work suggests that the implementation of the new ICRP phantoms and of the HAT model into the dose calculations will not change most organ and effective doses profoundly, but some differences will occur due to the closer proximity of organs and the more explicit treatment of electron transport implicit in the new ICRP phantoms.

Acknowledgements
This project has been supported by ICRP and the Swedish Radiation Safety Authority. It was partially carried out within the Collaborative Project "MADEIRA" (www.madeira-project.eu), cofounded by the European Commission through EURATOM Seventh Framework Programme (Grant Agreement FP7-212100).

References

[1] ICRP 1971 *Protection of the Patient in Radionuclide Investigations* ICRP Publ. 17 (Oxford: Pergamon Press)
[2] ICRP 1987 *Radiation Dose to Patients from Radiopharmaceuticals* ICRP Publ. 53 Ann. ICRP **18**(1-4)
[3] ICRP 1991 *Radiological Protection in Biomedical Research* ICRP Publ. 62 Ann. ICRP **22**(3)
[4] ICRP 1998 *Radiation Dose to Patients from Radiopharmaceuticals* Addendum 2 to ICRP Publication 53 ICRP Publ. 80 Ann. ICRP **28**(3)
[5] ICRP 2008 *Radiation Dose to Patients from Radiopharmaceuticals* A third amendment to ICRP Publ. 53 ICRP Publ. 106 Ann. ICRP **38**(1-2)
[6] ICRP 1979 Limits for Intakes of Radionuclides by Workers ICRP Publ. 30 Ann. ICRP **2**(3-4)
[7] ICRP 2002 *Basic Anatomical and Physiological Data for Use in Radiological Protection: Reference Values* ICRP Publ. 89 Ann. ICRP **31**(3-4)
[8] Snyder W S, Ford M R, Warner G G and Watson S B 1975 *MIRD* Pamphl 11 (New York, USA: Society of Nuclear Medicine)
[9] Cristy M and Eckerman K 1987 *Specific Absorbed Fractions of Energy at Various Ages from Internal Photon Sources* (Oak Ridge National Laboratory, Oak Ridge, TN) ORNL/TM-8381
[10] Eckerman K and Stabin M 2000 *Health Phys.* **78** 199-214
[11] ICRP 2007 The 2007 *Recommendations of the International Commission on Radiological Protection* ICRP Publ. 103 Ann. ICRP **37**(1-3)
[12] ICRP 1991 1990 *Recommendations of the International Commission on Radiological Protection* ICRP Publ. 60 Ann. ICRP **21**(1–3)
[13] ICRP 2009 *Adult Reference Computational Phantoms* ICRP Publ. 110 Ann. ICRP **39**(2)
[14] Leggett R W, Williams L R and Eckerman K F 1996 *A blood circulation model for reference man* Proc. of the 6th Int. Radiopharmaceutical Dosimetry Symp. (Oak Ridge, TN: Oak Ridge Associated Universities) ORISE 00-0164 487-499
[15] Stabin M G 2010 *Health Phys.* **99**(6) 811-813
[16] ICRP 2008 *Human Alimentary Tract Model for Radiological Protection* ICRP Publ. 100 Ann. ICRP **36**(1-2)

Journal of Physics: Conference Series **317** (2011) 012009 doi:10.1088/1742-6596/317/1/012009

Impact of low-dose CT scan in dual timepoint investigations: a phantom study

M A Micheelsen[1] and M Jensen[2]

[1] Department of Nuclear Medicine, Koege Hospital, Lykkebækvej 1, DK-4600 Koege, Denmark
[2] Hevesy Laboratory, Risoe-DTU, Technical University of Denmark, DK-4000 Roskilde, Denmark

E-mail: mamic@regionsjaelland.dk

Abstract. Dual timepoint FDG takeup investigations have a potential for separating malignant lymph nodes from non-malignant in certain cases of suspected lung cancer. One hour seems to be the optimal time interval between the two scans (50-120 min). Many of the new PET scanners benefit from image fusion with a CT image and also use the CT for attenuation correction. In any practical hospital setting, 1 hour is too long to occupy the scanner bed and a second CT procedure thus becomes necessary. This study tries to validate to what extent the dose/quality of the second CT scan can be lowered, without compromising attenuation correction, lesion detection and quantification. Using a standard NEMA phantom with the GE Discovery PET/CT scanner, taken in and out between scan sessions, we have tried to find the minimal CT dose necessary for the second scan while still reaching tissue activity quantification within predetermined error limits. For a hot sphere to background activity concentration ratio of 1:5, the average uptake (normalised by the time corrected input activity concentration) in a sphere of 6 cm^3 was found to be 0.90 ± 0.08 for the standard scan, yielding a dose of 5.5 mGy, and 0.90 ± 0.14 for a scan with lowest possible mAs product and lowest possible kV, yielding a dose of 0.65 mGy. With an insignificant increase in the uncertainty in the uptake measurement, we can get an order of magnitude reduction for the CT dose.

1. Introduction

While it is important to aspire toward as accurate diagnostics as possible, the benefits must be weighed against the potential hazards to patients and the costs it might impose. Dual timepoint FDG takeup investigations have a potential for separating malignant lymph nodes from non-malignant in certain cases of suspected lung cancer [1-3]. To detect a change in the FDG uptake of a lesion in two different scans, we must ensure that the uncertainty in a single measurement is smaller than the change we observe. This is done by eliminating or minimising any differences between the two scans. An optimal procedure would be to have the patient lying on the scanner bed - completely still - for the whole duration (scan-pause-scan). This would also have the benefit that the same CT scan could be used for the attenuation correction map for both scans, and the patient only receives the CT dose once. However, while patients referred for a lung PET scan are often quite uncomfortable and weak, lying completely still for 1 hour for each of the two scans, is not realistic. On the cost/benefit side, a small hospital with one PET scanner would have to drastically reduce the number of patients per day, if one procedure lasts 1.5 hours rather than 20 min in the scanner. So realistic implementation of a second

PET scan as a standard tool for diagnosing lung cancer would require a second CT scan, with the extra radiation dose to the patient this entails.

Current software has automated features that are capable of defining a boundary around lesions (areas of higher activity) and to determine the lesion uptake. This was used for volume detection as well as uptake determination.

The sole purpose of this study is to investigate if there is a significant effect on the uptake values found in a lesion in the PET reconstructed images, when different parameters for the CT scan are used for the attenuation map.

2. Materials and methods

The current phantom study was performed on a GE Discovery VCT 64 PET/CT scanner and data processing were carried out on a GE Advantage Workstation running SW version 4.4. The phantom was a standard NEMA phantom with 6 sphere inserts of various sizes. The overall phantom volume is 9.7 litres. "Hot" spheres and phantom background were loaded with known, activity calibrated F-18 solutions, prior to the scans. In between dual timepoint measurements, the phantom was kept stationary on the scanner bed. The scanner software was used to quantify volumes, average and max uptake values (kBq/ml) in that volume. Measured uptake values were compared to the correct values derived from the activity concentrations loaded into the phantom. The ratio between measured and calculated activity is used as a figure of merit and expressed as "correctness" (S). An S value of exactly 1 corresponds to a completely correct measurement. As scans and quantifications were repeated on a given phantom, embracing different activity levels due to decay, several S value determinations were performed for each sphere. For a dataset of such repeated determinations, the mean value, the standard deviation SD and the relative uncertainty RU (SD over mean) were calculated.

The aim of the study is to identify the impact of the CT dose on the correctness. However an initial series of experiments were performed to stage the relevance of this. In preparation, a series of runs were made. First the effect of "lesion" to "background" activity concentration ratio was investigated [5], and then the effect of "lesion" volume [4]. Neither should have an effect in a dual timepoint study, but will set a limit when intrinsic fluctuations become too large to trust a comparison between two measurements. The result from these studies showed a strong dependence but there is no simple way to define lower limits. As a reasonable baseline for this investigation, a signal to noise ratio of approximately 1:5 was chosen and a volume of 5.57 cm^3 was used for standard investigations to disregard volume effects. Few investigations with a 0.52 cm^3 volume were also conducted. For our purpose, these settings are in agreement with other studies [4, 5].

Second the dose was lowered successively by tuning the noise index, the mAs and the voltages. This was done to reveal any obvious bias or trends in the noise. Even though, there is some variability in the result (data not shown), it was not possible to deduce if these were tied to the quality of the CT scan or just intrinsic variations in the PET data. Thus, the next natural step was a more thorough comparison between two fixed sets of parameters, one that represented a standard patient scan (first scan, 120 kV and smartAmp feature) and a second set of parameters where we chose the lowest possible values (50 mAs and 80 kV). This change in parameters reduced the calculated CT dose to the phantom by a factor 8.4, from 5.47 to 0.65 mGy. The settings are referred as High Dose (HD) and Low Dose (LD).

To avoid bias on the datasets from decay, scans were made alternating between the two sets of parameters. An overview of these results from two different days can be seen in figure 1 and table 1.

3. Results

By far the largest uncertainties showed up in the recovery of the volume (data not shown). However, the uncertainty in the recovery of the uptake value is much smaller and it is surmised that a consistent determination of activity in a lesion can be made without demanding the same precision in volume detection.

In figure 1, the average correctness values are close to one, but with a tendency for deviations in activity concentration levels of 2 kBq/ml and below. This tendency is observed in both the HD and LD datasets and thus is not caused by the change in CT settings.

In table 1, the mean and standard deviation of the correctness value for all datasets as well as a combined set of HD and LD are shown. If there is a shift in the mean between the LD and the HD datasets, we expect to observe a remarkable increase in the relative uncertainty of the combined (HD+LD) dataset. If the datasets does have the same mean, the relative uncertainty becomes roughly: $0.5\ [(\sigma_{HD})^2 + (\sigma_{LD})^2]$. Hence, if the combined dataset has a relative uncertainty between the corresponding relative uncertainties of the individual sets, this means that the datasets are comparable. This is obvious for dataset 1 (table 1), but not so for dataset 2. To elaborate on this result, the relative differences between the means of the LD and HD correctness of the datasets were calculated and compared with the smallest relative uncertainty of the two sets, as shown in table 2. This gives a slightly more positive indication as the relative difference in the mean is quite small and even smaller than or comparable to the smallest fluctuation. Note that for all values, except the average value of the second dataset, HD has the smallest uncertainty.

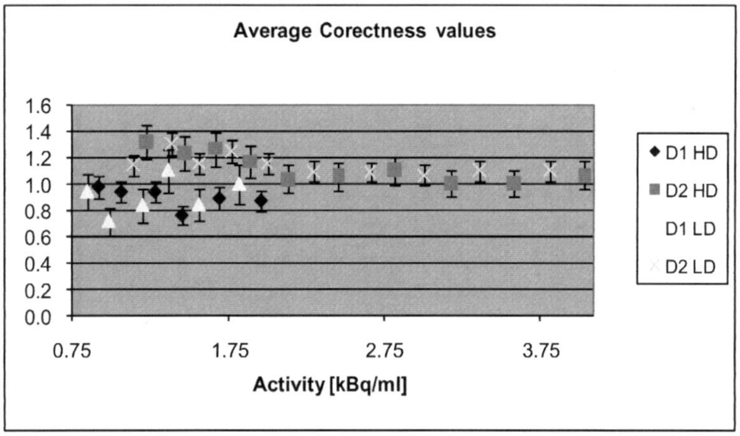

Figure 1: Average correctness value (S) as a function of the input activity concentration in a sphere of 5.57 cm^3. The 4 datasets are: the two different days (D1, D2) and the two CT parameter settings yielding Low Dose (LD) and High Dose (HD).

Table 1. The number of scans (N), the volume of spheres (V), the mean of the correctness value (S), uptake/input activity concentration in sphere, with the relative uncertainties.

	N	V (cm3)	$\langle S_{Max} \rangle$	$\sigma_{Max}/\langle S_{Max} \rangle$ %	$\langle S_{Av} \rangle$	$\sigma_{Av}/\langle S_{Av} \rangle$ %
Dataset 1: First scan activity ~ 2 kBq/ml, last scan activity ~ 0.7 kBq/ml, signal to noise ratio was 1:4.79						
HD [a]	6	5.57	1.53 ± 0.18	11.9	0.90 ± 0.08	8.7
LD [b]	6	5.57	1.49 ± 0.22	14.5	0.90 ± 0.14	15.4
Mixed [c]	12	5.57	1.51 ± 0.19	12.7	0.90 ± 0.11	11.9
Dataset 2: First scan activity ~ 4 kBq/ml, last scan activity ~ 1.1 kBq/ml, signal to noise ratio 1:4.77						
HD	10	0.52	0.86 ± 0.15	18	0.53 ± 0.10	19.7
LD	10	0.52	0.90 ± 0.16	17.2	0.55 ± 0.13	23.9
Mixed	20	0.52	0.87 ± 0.15	17.2	0.53 ± 0.11	21.6
HD	10	5.57	1.68 ± 0.13	7.8	1.09 ± 0.09	8.2
LD	10	5.57	1.81 ± 0.19	10.5	1.15 ± 0.08	6.6
Mixed	20	5.57	1.78 ± 0.20	11.1	1.14 ± 0.09	8.3

[a] High Dose
[b] Low Dose
[c] Combination of LD and HD

The relative uncertainty is around 10% for the 5.57 cm^3 volume with a tendency to be slightly larger for the LD dataset. The small volume gives rise to a much larger relative uncertainty (~20%). This is important since the relative uncertainty will set a limit to the acceptance of an important increase in lesion. Other studies have suggested 2 ml as a minimal volume for accurate uptake measurement [5].

Table 2. The relative difference, $RF = (\langle S_{HD} \rangle - \langle S_{LD} \rangle)/\langle S_{HD} \rangle$, between the mean S values of the LD and HD for the two datasets (D1 and D2) and the smallest relative uncertainties (RU).

	D1		D2	
	RF %	RU %	RF %	RU %
Av.	0.7	8.7	6.9	6.6
Max.	2.3	11.9	5.0	7.8

4. Conclusion and discussion

The aim of this study has been to uncover any change or bias in correct retrieval of uptake values between two scans that differ only by a change in the parameters for the CT scan used for attenuation correction. Under the assumptions of the performed experiments, the conclusion is that there is no clinically important difference. The results do reveal generally large fluctuations within the datasets, and these should be used to help in determining how much the uptake needs to change before we consider any change significant. With a lesion activity concentration of 4 kBq/ml or less and a background to noise ratio of roughly 1:5 any change less than 10% can not be considered significant. In the actual human scanning situation, with more scatter, repositioning problems and movements, the limits to what can be considered significant is probably even higher.

The average and maximum values are fairly consistent; although the variability is lower in the average values. The relative uncertainty in the average uptake in the large sphere is 10%, so in this case increases in uptake of more than 10% could be detected with certainty. Initial patient studies on site have shown a tendency for the investigated lesions to have quite a large relative increase in uptake. Thus, at present 41 lesions from 17 consecutive patients have been investigated and for these, 70% (± 14%, 95% confidence) had an increased uptake of more than 10% (with the second scan on average 85 ± 6 min after the first). None of these lesions have yet been verified or rejected as malignant, but they reveal that a large quantity of the data we work with, have an increase in uptake that is significant (e.g. above 10%). Other studies show similar tendencies [1], 20 malignant lesions showed an average increase of 20.5%, which in our study would be considered adequate. Lesion sizes were not considered for these data.

The conclusion concerning a lower boundary for the significance is only valid in the case of a well calibrated scanner. If the dual timepoint method was implemented as standard procedur, it would be advisable to test this as a part of the standard QA. In this study, we used two different voltage settings for the HD and LD scans, from 120 kV to 80 kV. By only changing the mAs and maintaining the same kV the dose will increase (to 1.7 mGy); any possible systematic errors have been removed.

Acknowledgements
Patient data were provided by courtesy of Chief Physician, MD Anna Buhl. The authors would also like to thank technician Birgitte Høj for valuable assistance in the scan sessions and Chief Physician, MD Knud Heine Stokholm for valuable comments and additions to the manuscript.

References
[1] Matthies A, Hickeson M, Cuchiara A and Alavi A 2002 *J. Nucl. Med.* **43** 871-5
[2] Basu S and Alavi A 2007 *Eur. J. Nucl. Med. Mol. Imaging* **34** 1527-9
[3] Uesaka D *et al* 2008 *J. Nucl. Med.* **49** 1606-12
[4] Tylski P *et al* 2010 *J. Nucl. Med.* **51** 268-76
[5] Daisne J-F *et al* 2003 *Radiotherapy and Oncology* **69** 247-9

Safety of knee radiosynovectomy with yttrium - 90

M Kempińska[1], P Lass[1], J B Ćwikła[2] and P Żbikowski[3]

[1] Department of Nuclear Medicine, Medical University of Gdańsk, 7 Dębinki Str., 80-211 Gdańsk, Poland
[2] Central Clinical Hospital of the Ministry of Internal Affairs and Administration in Warsaw, Department of Nuclear Medicine, Poland
[3] Central Clinical Hospital of the Ministry of Internal Affairs and Administration in Warsaw, Orthopedics and Traumatology Clinic, Poland

E-mail: mkempinska@gumed.edu.pl

Abstract. Radioisotope knee synovectomy is based on an Yttrium - 90 citrate injection (185 - 222 MBq) into the knee joint cavity. The performance of procedure needs participation of a nuclear medicine specialist as well as an orthopedist or a rheumatologist and a technologist, who prepares radiopharmaceuticals. The ionization doses for patients and personnel depend not only on the injected activity, but also on the method and process of injection and the radioactivity measurement procedure used. The aim of this study is the evaluation of the degree of radiation exposure of patients and medical personnel during the performance of therapy with ^{90}Y.

1. Introduction

According to ICRP classification, there are three types of radiation exposure risks, i.e. medical exposure, occupational exposure of medical personnel and finally general public exposure [1]. In case of radiation exposure it is necessary to define the emission properties of the isotope used. ^{90}Y is a β-emitter, so it induces secondary radiation called Bremsstrahlung also during the radiopharmaceutical's preparation.

Pure β-ray emitters are used as therapeutic radionuclides, because of the short range of radiation (typically less than 1 cm) in the tissue. It has been implicitly assumed that there is no significant external radiation hazard from internally emitted β-rays. With the relatively low energy (of the order of 100 keV or less) of β-rays typically encountered in nuclear medicine and the low effective atomic number (~7.9 [2, 3]) of soft tissue, there is virtually no bremsstrahlung produced in vivo, and the assumption of no external radiation hazard from β-rays is altogether reasonable. With increasing therapeutic use of relatively high-energy emitted β-rays (of the order of 1 MeV or more, for example ^{90}Y with maximum energy of 2.27 MeV [4]) particularly in materials with high atomic numbers such as bone with an effective atomic number of ~ 20, equivalent to that of calcium,, production in vivo of bremsstrahlung is sufficient for external detection and imaging [5].

The most frequently used method of measurement is the thermoluminescence dosimetry. This method may give contradictory results due to the high level of uncertainty (read error). In some cases (Bremsstrahlung or β/γ - emitters), gamma camera measurements seem to be useful and precise tool in dosimetry.

2. Materials and methods

Dosimetry measurements were performed in the Central Clinical Hospital of the Ministry of Internal Affairs and Administration in Warsaw. Patients were suffering from diseases like rheumatoid arthritis, psoriatic arthropathy, serosynovitis and haemophilic arthropathy.

2.1. Thermoluminescence dosimeters

For the quantification of the exposure to ionizing radiation thermoluminescence dosimeters were used. The personal dose, equivalent at 0.07 mm and 10 mm depth in the body of 70 persons and 30 medical personnel, and the kinetic energy released per unit air mass were calculated.
The thermoluminescence dosimeters that were used to determine the:

- personal dose equivalent at 0.07 mm depth in the body Hp(0.07), in mSv, are dosimeter type PI-01,
- personal dose equivalent at 10 mm depth in the body Hp(10), in mSv, are dosimeter type DI-02,
- kinetic energy released per unit air mass (air KERMA) *Ka*, in mGy, are dosimeter type DS-04.

Measurements of dosimeters PI-01, DI-02 and DS-04 were performed at the Laboratory of Individual and Environmental Dosimetry of the Institute of Nuclear Physics in Kraków, Poland.

Each patient taking part in this study was equipped with a dosimeter type: PI-01 (attached immediately after injection close to the place of injection), DI-02 (placed under clothing on the patient's chest) and DS-04 (0.5 m to 1 m over the patient's bed). Exposure of the activity was measured for a minimum of 48 h. Special instructions had been prepared for patients to explain how to deal with dosimeters and a questionnaire helped to understand the process of dosimetry exposure (patient's compliance). Additionally, all medical records were collected for each patient.

2.2. Scintigraphy

For proper assessment of radiation emitted by a radionuclide, its energy spectrum and reduction of radiation were established using gamma camera analysers. Static studies (10 min) of patient's knee joint (20 knee joint) were performed. Imaging studies were done using the dual head gamma camera Ecam (Siemens, Erlangen, Germany). Whole body study after injection was performed in 6 patients for the purpose of assessment of the extra-articular location of the tracer.

3. Results

The average number of radiosynovectomy procedures performed per month in each Polish Nuclear Medicine department is six. Limit dose equivalent values are defined by the Atomic Law and the safety of RS experiments is met [6]. Individual dosimetry of the medical personnel proved that the small number of conducted procedures is safe for the medical staff (table 1).
Similarly the absorbed dose to the patients was within the acceptable limits, if good radiation protection practice is implemented (table 2).
Yttrium-90 citrate colloid is a radiopharmaceutical that usually does not remain stable in the target organ. Qualitative analysis had shown an absence of the radiopharmaceutical flow outside the joint cavity (figure 1).

Dosimetric procedures proved that its local concentration is changing in time, so that compartmental analysis can be implemented to check the influence of the disease pathophysiology on the kinetics of the radiopharmaceutical (figure 2).

Table 1. Mean values of the dosimeter readouts (TLDs) for medical personnel; values have been normalized to 1 MBq of injected radiopharmaceutical activity.

TLDs	Medical personnel (μSv)	
	Physician (15 persons)	Technologist (15 persons)
PI-01-Hp(0.07)	1	3
DI-02-Hp(10)	0.8	1
DS-04-*Ka*	0.7	0.9

Table 2. Mean values of the dosimeter readouts (TLDs) for patients; values have been normalized to 1 MBq of injected radiopharmaceutical activity.

TLDs	Dose (μSv)
PI-01 – Hp(0.07)	68
DI-02 – Hp(10)	0.7
DS-04 – *Ka*	0.8

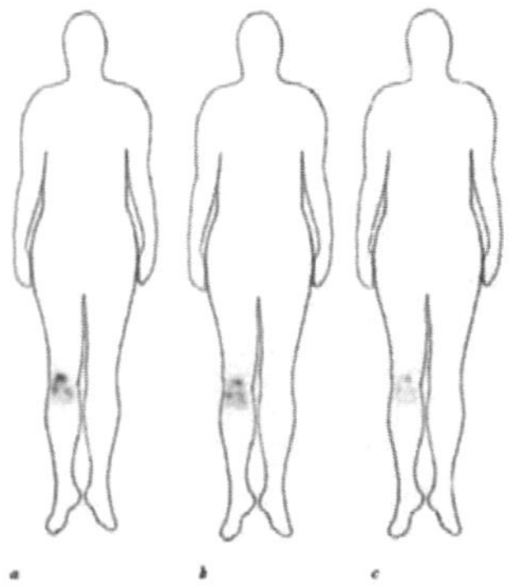

Figure 1. Whole-body scans after (a) 1 h, (b) 24 h, (c) 96 h after injection of the radiopharmaceutical.

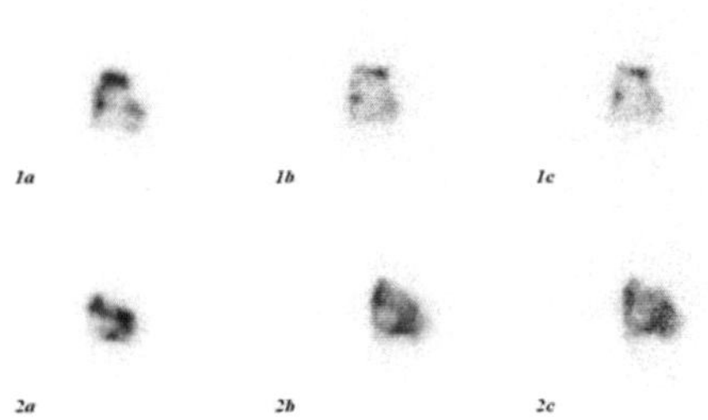

Figure 2. Static scans of the joint after (a) 1 h, (b) 24 h, (c) 96 h after injection of the radiopharmaceutical.

4. Conclusion

Scintigraphy of the knee joint at a given moment during the therapy provides a useful means of monitoring the accumulation process of radiopharmaceutical. It gives very useful information for a physician. Using this information, the physician can foresee the results of the undertaken therapy. The analysis of the whole body scintigraphy images proved that no more than 10% of the original radiopharmaceutical activity injected to the joint escaped outside the joint cavity. The biodistribution of radiopharmaceuticals in the patient's body requires more dosimetric experiments.

References

[1] Filipiak B, Mlicki K and Haratym Z 1999 Should internal risk of medical personnel be monitoring? *Współcz Onkokl* **6** 259–61 (in Polish)
[2] Johns H E and Cunningham J R 1969 *The Physics of Radiology*, 3rd ed. (IL: Charles C. Thomas, Springfield) p 210
[3] White D R Woodard H Q and Hammond S M 1987 Average soft-tissue and bone models for use in radiation dosimetry *Br. Radiol.* **60** 907-913
[4] EANM Procedure Guidelines for Radiosynovectomy 2003 *Eur. J. Nucl. Med.* **30**(3) BP12-16
[5] Zanzonico B, Binkert B and Goldsmith S 1999 Bremsstrahlung radiation exposure from pure β-ray emitters *J. Nucl. Med.* **40** 1024–8
[6] Decree of the Council of Ministers from 18.01.2005 on the limits of ionizing doses 2006 *Dziennik Ustaw* no. **20** item 168 (in Polish)

Dosimetry of exendin-4 based radiotracer for glucagonlike peptide-1 receptor imaging: an initial report

M Tomaszuk[1], A Sowa-Staszczak[1], W Lenda-Tracz[1], B Glowa[1], D Pach[1], M Buziak-Bereza[1], A Stefanska[1], B Janota[2], D Pawlak[2], R Mikolajczak[2] and A B Hubalewska-Dydejczyk[1]

[1] Nuclear Medicine Unit, Department of Endocrinology, Krakow University Hospital, Kopernika 17, 31-501 Krakow, Poland
[2] Institute of Atomic Energy, Radioisotope Center POLATOM, 05-400 Otwock-Swicrk, Poland

E-mails: alahub@cm-uj.krakow.pl, mtomaszuk@su.krakow.pl

Abstract. Overexpression of glucagonlike peptide-1 (GLP-1) receptors in human tumours is a potential target for future imaging and therapy. The GLP-1 receptor imaging using [Lys^{40}(Ahx-HYNIC-^{99m}Tc/EDDA)NH$_2$]-exendin-4 could be useful in the localization of unknown insulinoma focus. The aim of this study was to present the first experience of our unit with the new radiopharmaceutical and its dose estimates. Imaging studies and dose assessment, according to the MIRD schema and MIRD Pamphlet No.11, were performed for 3 patients (2 with suspicion of insulinoma, 1 with suspected insulinoma recurrence). In the first case suspicion of insulinoma was not confirmed. In the second case localized accumulation of tracer in the pancreas was removed by surgery and the clinical symptoms of insulinoma receded. In the third case, pathological accumulation of tracer was localized and recurrence of insulinoma was confirmed in fusion with CT images. The biological half-time did not exceed 2.7.h. The effective half-time did not exceed 4.8 h. The total-body radiation dose did not exceed 0.0038 mGy/MBq and is comparable with the radiation dose to patient after somatostatin receptor scintigraphy. The highest radiation dose was calculated for kidneys (~ 0.070 mGy/MBq). [Lys^{40}(Ahx-HYNIC-^{99m}Tc/EDDA)NH$_2$]-exendin-4 is a good candidate for clinical GLP-1 receptor imaging studies and appears safe for the patient from radiological safety point of view.

1. Introduction

During the past two decades, peptide receptor scintigraphy with the use of somatostatin analogs (SRS, somatostatin receptor scintigraphy) has become an efficient and clinically accepted method for visualization of receptor-positive human tumours and their metastasis. But not all of the tumours overexpressed somatostatin receptors. Less than 60% of insulinomas expressed somatostatin receptor subtype 2.The sensitivity of the Octreoscan is assessed below 50% respectively [1-3].

Very high incidence and density of another regulatory peptide receptor were reported in insulinomas' cell membrane [2]. Overexpression of glucagon-like peptide 1 (GLP-1) receptors in insulinomas may be used for effective localization of the change and consequently may result in successful surgical treatment, which is the only method for potentially relieving patients from the disease. The first tests showed that GLP-1 receptor imaging using [Lys^{40}(Ahx-DTPA-^{111}In)NH^2]-exendin-4 could be useful in the non invasive localization of insulinomas [4-5]. Compared with ^{111}In-labelled agents, examinations performed with use of ^{99m}Tc-labelled radiopharmaceuticals are characterized by many procedural advantages related to this isotope's physical properties. The new tracer - [Lys^{40}(Ahx-HYNIC-^{99m}Tc/EDDA)NH_2]-exendin-4, may improve the quality of images and radiation safety for patients and staff.

It is important to have information about the dose assessment from this new clinically useful new imaging option. Up to date the biokinetic characteristics and dose estimates have been reported only for animal studies.

The aim of this study is to present the first experience of our unit with this new radiopharmaceutical and our effort to establish the biokinetic characteristics of [Lys^{40}(Ahx-HYNIC-^{99m}Tc/EDDA)NH_2]-exendin-4, and to estimate the radiation dose to patients with suspicion of or confirmed insulinoma.

2. Materials and methods

2.1. Materials

Imaging studies and dose assessment were performed for 3 patients (3 females, 52 ± 6 y): (a) 48-years-old woman with hypoglycaemias, no lesion in CT and no uptake in SRS, (b) 57-years-old women with severe clinical hypoglycaemias, no lesion in CT and no uptake in SRS, and (c) 62-years-old women with malignant insulinoma after surgery with metastases to the liver in CT, with suspected recurrence and no pathological uptake in SRS. All three patients gave their written consent. The study was approved by the local Medical College Ethics Committee. The average injected activity of [Lys^{40}(Ahx-HYNIC-^{99m}Tc/EDDA)NH_2]-exendin-4 was 740 MBq.

2.2. *[Lys^{40}(Ahx-HYNIC-^{99m}Tc/EDDA)NH_2]-exendin-4 preparation and radiochemical purity*

Technetium-99m labeled [Lys^{40}(Ahx-HYNIC/EDDA)NH_2]-exendin-4 was obtain from lyophilized kits prepared by the Institute of Atomic Energy, Radioisotope Center POLATOM. Exendin-4 was modified C-terminally with Lys^{40}-NH_2, where the lysine side chain was conjugated with Ahx-HYNIC (Ahx is aminohexanoic acid). Tricine and EDDA as co-ligand for ^{99m}Tc were involved. The radiopharmaceutical procedure was carried out in the Nuclear Medicine Unit of the Endocrinology Department, Cracow University Hospital and was performed under aseptic conditions. Two-vials freeze-dried kits were used for radiolabelling with 0.3-1.5 ml pf ^{99}Mo/^{99m}Tc generator eluate (10-50mCi) followed by 20 min incubation at 80 ^{0}C. The TLC method was used for assessing the radiochemical purity of the compound.

2.3. Imaging acquisition

All images were acquired with a dual-head, large field of view E.CAM gamma camera (Siemens, 2000) with low-energy high resolution (LEHR) collimators. Whole-body scans were performed 1, 2, 3, 6, 24 and 30 h using the same procedure. The camera settings were as follows: 1024 x 256 matrix, 12 cm/min scan speed, dual-energy window for scatter correction. The standard of known activity (vial with about 10 MBq of ^{99m}Tc) was used to obtain the imaging system calibration factor. A standard was always scanned with the patient in the same position. To obtain the transmission factor through each source region of interest (ROI), a whole-body scan of the patient was performed with identical settings and with a ^{57}Co flood source.

2.4. Radiopharmaceutical biokinetics and dose assessment

The external conjugate view counting pair method was used for analyzing whole-body images.

Firstly, ROIs were drawn visually over the source organs (kidneys, liver and lungs) on the whole body 1 h anterior scan image. These were flipped horizontally on to the posterior scan. Secondly, the ROIs were copied on to the rest of the images. The counts under the ROIs were converted to activity values after the subtraction of the background, the previously determine camera detection efficiency and attenuation correction, according to the MIRD pamphlet No. 16 equation [6-7].

All time-activity data were expressed as a percentage of the administered activity (% I.A. – the percent injected activity). For the whole-body data the activity at t = 0 (100% I.A.) were estimated by fitting a mono-exponential function to the first two data points. For the source-organs, the activity at t = 0 was set to zero, assuming the non-instantaneous uptake in these organs.

Assuming first-order kinetics of the tracer, the area under the time-activity curve for the whole-body (the time-integrated activity coefficient), the biological half-time and the effective half-time for tracer were estimated by fitting a mono-exponential function to the data points. Areas under the time activity curves for 3 source organs (kidneys, liver and lungs) were generally estimated using the trapezoid method. A mono-exponential function was fitted to the two last data points for a better estimation of the area under the time-activity curve after the last acquired image (after 30 h).

The absorbed dose coefficient $d(r_T,T_D)$ was evaluated according to the Medical Internal Radiation Dose (MIRD) system with S values according to the MIRD Pamphlet No.11 [8-9]. The total effective dose, using weighting factors defined in ICRP Publication 103, was also evaluated [10].

3. Results

The average radiochemical purity of the administered compound, prepared according to manufacturers' instruction and determined by TLC, was calculated to be around the 90% level.

In the first case (48-years-old woman), suspicion of insulinoma was not confirmed. In the second case (57-years-old women), accumulation of tracer in pancreas localization was remove by surgery. The postsurgical resolution of all the symptoms was observed. In the third case (62-years-old women), pathological accumulation of tracer was localized and the recurrence of insulinoma was confirmed in the CT fussed image.

The percentage of the injected activity after 1 h, 6 h and 24 h in the whole-body and source organs are shown in table 1. Also in table 1 the biological half-time and the effective half-time of the tracer estimated for each patient are shown. Absorbed dose estimates are shown in table 2.

Table 1. The percentage of the injected activity after 1 h, 6 h and 24 h in the whole-body and source organs, the biological half-time and the effective half-time of the tracer.

	Patient 1			Patient 2			Patient 3		
Time (h)	1	6	24	1	6	24	1	6	24
Whole-body (% I.A.)	89.3	36.7	4.0	87.5	23.6	2.5	77.7	25.0	1.6
Kidneys (% I.A.)	12.1	8.4	1.4	12.7	8.7	1.1	14.1	7.3	1.7
Liver (% I.A.)	5.3	2.2	0.2	4.5	1.7	0.2	2.8	0.8	0.1
Lungs (% I.A.)	5.8	1.6	0.1	4.5	0.9	0.1	4.5	0.8	0.1
Biological half-time (h)	2.7			2.6			2.5		
Effective half-time (h)	4.8			4.6			4.3		

4. Discussion

Imaging with GLP-1 receptor agonists is a novel option for localization of tumours overexpressing GLP-1 receptors. First clinical studies using ^{111}In-labeled GLP-1 receptor agonist, [Lys^{40}(AhxDOTA-^{111}In)NH_2]-exendin-4, revealed high sensitivity. Six out of six benign insulinomas were detected [5]. Since then, a new compound Lys^{40}(Ahx-HYNIC-^{99m}Tc/EDDA)NH_2]-exendin-4 was synthesized.

In this paper the biokinetics and dose assessment of Lys^{40}(Ahx-HYNIC-^{99m}Tc/EDDA)NH_2]-exendin-4 for SPECT/CT studies are presented.

[Lys^{40}(Ahx-HYNIC-^{99m}Tc/EDDA)NH_2]-exendin-4 showed mainly renal clearance. Activity was mainly accumulated in the kidneys, and to a lesser extends in the liver and lungs. The biological half-time did not exceed 2.7 h. The effective half-time did not exceed 4.8 h. The whole-body radiation dose did not exceed 0.0038 mGy/MBq. Table 2 shows that the highest radiation dose based on this data was estimated for the kidneys (~ 0.070 mGy/MBq). The effective dose calculated by these dose estimates did not exceed 0.0023 mSv/MBq and is comparable with the radiation dose to the patient from SRS performed with ^{99m}Tc-EDDA/HYNIC-Tyr^3-octreotide [11].

Table 2. Organ absorbed dose estimates in mSv/MBq for [Lys^{40}(Ahx-HYNIC-^{99m}Tc/EDDA)NH_2]-exendin-4.

	Patient 1	Patient 2	Patient 3
Kidneys	0.0660	0.0813	0.0768
Liver	0.0057	0.0060	0.0038
Lungs	0.0049	0.0037	0.0033
Spleen	0.0032	0.0039	0.0036
Pancreas	0.0029	0.0034	0.0030
Intestine	0.0042	0.0050	0.0050
Red marrow	0.0016	0.0019	0.0017
Thyroid	0.0001	0.0001	0.0001
Muscles	0.0006	0.0007	0.0007
Ovaries	0.0004	0.0005	0.0005
Total Body	0.0038	0.0032	0.0030
Effective dose	0.0023	0.0022	0.0021

Wild *et al* [12] evaluated the pharmacokinetics and radiation dose after injection of [Lys^{40}(Ahx-HYNIC-^{99m}Tc/EDDA)NH_2]-exendin-4 in animals' studies. For all calculations, the authors made the assumption that the Rip1Tag2 mouse biodistribution, determined as the % IA / organ, was the same as the human biodistribution. Our study showed faster retention of the tracer in human organs and the radiation dose and the effective dose to the other organs were at similar levels.

It should be emphasized that values estimated in this study are not used to evaluate the risk to a given individuals. This initial report is based on data from only 3 patients. Data from a larger number of patients is required for a complete assessment of the radiation safety of the patient. It is also necessary to take into account the blood activity and excretion concentrations. This paper provides only preliminary evaluation of organ doses from ^{99m}Tc labeled GLP-1 receptor imaging. The data should the considered only as the basis for further evaluation of radiation safety of the radiopharmaceutical.

5. Conclusions

[Lys^{40}(Ahx-HYNIC-^{99m}Tc/EDDA)NH_2]-exendin-4 is a good candidate for clinical GLP-1 receptor imaging studies and appears to be safe for the patient from a radiological safety point of view. The results provide the basis for further evaluation of the radiation safety of the radiopharmaceutical.

Acknowledgements
This study was supported by the Polish State Committee for Scientific Research (N N402 445039).

References

[1] Zimmer T *et al* 1996 *Gut* **39** 562-8
[2] Reubi J C and Waser B 2003 *Eur. J. Nucl. Med. Mol. Imaging* **30** 781-93
[3] Grant C S 2005 *Best Pract. Res. Clin. Gastroenterol* **19** 783-98
[4] Wild D *et al* 2008 *N. Engl. J. Med.* **359** 766-8
[5] Christ E *et al* 2009 *J. Clin. Endocrinol. Metab.* **94** 4398-405
[6] Siegel J A *et al* 1999 MIRD Pamphlet No.16 *J. Nucl. Med.* **40** 37S-61S
[7] Stabin M G 2008 *Fundamentals of Nuclear Medicine Dosimetry* Springer, LLC
[8] Bloch W E *et al* 2009 MIRD Pamphlet No.21 *J. Nucl. Med.* **50** 477-84
[9] Synder W S *et al* 1975 *MIRD* Pamphlet No.11 (New York: Society of Nuclear Medicine)
[10] ICRP 2007 Publ. 103, *Ann. ICRP* **37** 1-332
[11] Gonzalez-Vazquez A *et al* 2006 Biokinetica *App. Radiation Isotop.* **64** 792-7
[12] Wild D *et al* 2010 *J. Nucl. Med.* **51** 1059-67

Journal of Physics: Conference Series **317** (2011) 012012 doi:10.1088/1742-6596/317/1/012012

Effective doses to family members of patients treated with radioiodine-131

M Zdraveska Kocovska[1], O Vaskova[1], V Majstorov[1], S Kuzmanovska[1], D Pop Gjorceva[1] and V Spasic Jokic[2]

[1] Institute of Pathophysiology and Nuclear Medicine, Medical Faculty, University of "Sts Cyril and Methodius", Skopje, Former Yugoslav Republic of Macedonia
[2] Technical Faculties, University of Novi Sad, Serbia

E-mail: mzk2003@hotmail.com

Abstract. The purpose of this study was to evaluate the effective dose to family members of thyroid cancer and hyperthyroid patients treated with radioiodine-131, and also to compare the results with dose constraints proposed by the International Commission of Radiological Protection (ICRP) and the Basic Safety Standards (BSS) of the International Atomic Energy Agency (IAEA). For the estimation of the effective doses, sixty family members of sixty patients, treated with radioiodine-131, and thermoluminiscent dosimeters (Model TLD 100) were used. Thyroid cancer patients were hospitalized for three days, while hyperthyroid patients were treated on out-patient basis. The family members wore TLD in front of the torso for seven days. The radiation doses to family members of thyroid cancer patients were well below the recommended dose constraint of 1 mSv. The mean value of effective dose was 0.21 mSv (min 0.02 - max 0.51 mSv). Effective doses, higher than 1 mSv, were detected for 11 family members of hyperthyroid patients. The mean value of effective dose of family members of hyperthyroid patients was 0.87 mSv (min 0.12 - max 6.79). The estimated effective doses to family members of hyperthyroid patients were higher than the effective doses to family members of thyroid carcinoma patients. These findings may be considered when establishing new national guidelines concerning radiation protection and release of patients after a treatment with radioiodine therapy.

1. Introduction

Many types of cancer and some other non-malignant diseases can be treated with radiations emitted by radionuclides. The unsealed radionuclides, that are injected, ingested, or inhaled, and which move through the body, are referring to as radiopharmaceuticals. These can localize in body tissues until they decay or they can be eliminated through various pathways, such as sweat and saliva and excreted into urine and feces. The radionuclides used for radiopharmaceutical therapy are usually relatively short-lived beta emitters. Most of these radionuclides also emit photons, which usually contribute minimally to the treatment dose, but produce an undesirable radiation field emanating from the patient.

The most frequently used radiopharmaceutical for the treatment of thyroid diseases, such as Thyroid Cancer and Hyperthyroidism, is radioactive iodine-131. It has very high success rate in the treatment of patients with thyroid diseases and it has also proven to be safe and a relatively inexpensive treatment. The treatment renders the patient radioactive. The patients, treated with radioactive iodine-131, present a radiation hazard to other individuals such as hospital staff, the patient's family and members of the public with whom a treated patient may come in close contact. This situation can be overcome by imposing restriction on the behavior of the patient, to minimize the dose to close relatives and other individuals. In 1991, the International Commission of Radiological Protection (ICRP) [1] has recommended a radiation constraint of 1 mSv/y to the general population. According to the Basic Safety Standards Directive (BSS) [2], the dose limits to the general public are not valid for "exposure of individuals, who are knowingly and willingly helping, other than as part of their occupation, in the support and comfort of in-patient and out-patients undergoing medical diagnosis or treatment". Proposed dose constraint from the BSS is: 0.3 mSv per episode for public, 1 mSv for children, 3 mSv for adults up to sixty years of age, and 15 mSv for adults more than 60 years old.

The implementation of this guideline differs between countries. In the Institute of Pathophysiology and Nuclear Medicine, Skopje, about 50 thyroid cancer patients are treated on in-patient basis and, approximately, the same number of hyperthyroid patients are treated, ambulatory, on out-patient basis.

According to the local hospital rule and the old guidelines, the maximum given activity to hyperthyroid patients, treated on out-patients basis, is 1110 MBq. The new, not yet established guidelines, proposed to reduce the maximum given activity to hyperthyroid patients, treated on out-patient basis, to 555 MBq. This group of patients presents great radiation hazard to their family members. Upon discharge from hospital, the patients as well as their family members are given brief radiation safety instructions.

The aim is to minimize the transfer of radioactive material to persons coming in close contact with the patient. There are several papers in the literature concerning the subject of doses received by family members of thyroid cancer and hyperthyroid patients [3-8]. Most of the published studies agree that doses to the family members are bellow the proposed dose constraint of 1 mSv. But there are also several papers [5, 9] that present cases where children or other persons have received higher radiation doses than the proposed dose limit and, usually, this is the case with hyperthyroid patients and their relatives.

2. Purpose

The main purpose of this study was to estimate the radiation exposure to family members of patients treated with radioiodine-131 either for thyroid carcinoma or hyperthyroidism at our nuclear medicine centre. The other purpose was to use the results to identify necessary restrictions to ensure that the recommended dose constraint proposed by ICRP and the BSS of IAEA are met.

3. Materials and methods

The study population comprised thirty family members of the same number of thyroid cancer patients and thirty family members of thirty hyperthyroid patients. The administered dose for the treatment of thyroid cancer patients ranged from 3700 MBq to 5550 MBq of radioiodine-131. Mean administered dose was 3539 MBq. Twenty six patients received 3700 MBq, two patients received 4440 MBq and two patients received 5550 MBq.

They were hospitalized for three days after administration and their dose rate was measured every day at a distance of 0.25 m, 0.5 m, 1.0 m and 2.0 m by a medical physicist. The dose rate measurements were performed with a calibrated survey meter "mini-rad" series 1000, Morgan. When the level of 8 μSv/h at 2 m was reached, the patients were released from hospital. Upon discharge, the patients were given radiation safety instructions for their further behavior, in order to minimize the transfer of radiation to persons coming in close contact with them, especially children and pregnant women. Their relatives wore the TLD for one week and they were informed not to stay very close to the patient; and if so, to reduce the time of staying. It was suggested to maintain the distance between them and patient more than 2 m and to reduce time of staying to less than 10 minutes up to one hour.

Hyperthyroid patients were treated with 185 MBq to 1295 MBq of radioiodine-131. Mean administered dose activity was 683 MBq. External dose rate measurements were performed at the same distance as thyroid cancer patients, fifteen minute after administration of therapy. Afterwards, they were released from hospital.

All the patients were interviewed and were informed about the research aims by a medical physicist and a physician. They signed an agreement for participating in the study. Family members groups consisted of 12 females and 18 males in the hyperthyroid group and 24 males and 6 females in the thyroid cancer group. Their age varied from 15 up to 80 years. The effective dose measurements were carried out with thermoluminiscent dosimeters, model TLD 100, which contain 3 mm^2 hot pressed chips of lithium fluoride (LiF:Mg,Ti), encapsulated between two sheets of Teflon 10 mg/cm^2 thick and mounted on an aluminium substrate with-bar code and within shielded filter holders (type 8814 Harshaw). The detection threshold of the dosimetry system is 0.0054 mSv. TLD's were most appropriate to estimate radiation because the amount of ionizing radiation is directly proportional to the effective dose [10]. Actually, the depth dose Hp (10) was estimated. These types of dosimeters have photon energy response for gamma rays that ranges from 15 keV - 3 MeV (IEC 1066). The TLD Reader and Cards are calibrating on a regular basis. The combined standard uncertainty of a dosimetry system is less than 15%. Control TLDs were kept separately to measure the background. The background readings were subtracted from the dose readings of the relatives.

4. Results and discussion

4.1 Thyroid cancer patients

Table 1 presents the effective dose to relatives of the thyroid cancer patients treated with radioiodine-131 and measured with TLD dosimeters. The TLDs were worn by the relatives for one week. The range of the effective dose varied from 0.02 mSv to 0.51 mSv. For three family members, the TLD showed a value of 0 mSv and the explanation was that patients continue to be isolated from others after they had left the hospital.

Table 1. Effective doses (EF) to family members of thyroid cancer patients.

No.	EF (mSv)	Sex	Age (y)	No.	EF (mSv)	Sex	Age (y)
1	0.05	m	28	16	0.40	m	40
2	0.00	m	35	17	0.51	m	70
3	0.00	m	26	18	0.08	m	40
4	0.27	f	38	19	0.43	m	40
5	0.00	m	46	20	0.06	m	41
6	0.16	m	35	21	0.33	f	60
7	0.02	f	50	22	0.02	m	41
8	0.43	m	35	23	0.15	m	61
9	0.32	f	33	24	0.20	m	65
10	0.23	f	73	25	0.41	m	40
11	0.31	m	32	26	0.17	m	38
12	0.25	m	70	27	0.23	m	56
13	0.17	f	60	28	0.37	m	60
14	0.20	m	44	29	0.02	m	58
15	0.15	m	40	30	0.35	m	48
				Mean	**0.21**		

The mean value of the effective dose to relatives of thyroid cancer patients is 0.21 mSv. Figure 1 presents the effective doses to relatives of thyroid cancer patients versus proposed dose constraint of 3 mSv for adults and 15 mSv for older than 60 years of age according to the BSS. The family members' age varied from 28 years up to 73 years. The results show that the effective doses to all family members were well below the recommended dose constraints of 1 mSv (recommended dose for children by the BSS).

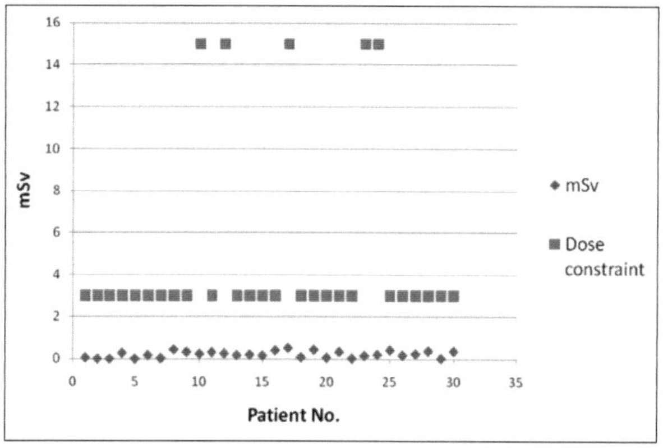

Figure 1. Effective doses to the relatives of thyroid cancer patients and the dose constraints of 3 mSv for adults and 15 mSv for older than 60 years of age, according to the BSS.

The results agreed with the previously published paper by Buchan *et al* [11] and proved that the protocol used for performing radioiodine therapy for thyroid cancer patient is sufficiently safe. Three days of hospitalization and dose rate measurements should be continued. It is recommended to ask patients to sleep alone for additional seven days and avoid close contact with other people, children and pregnant women [11]. Providing written instructions to patients about their behavior at home will improve the process of optimizing radiation protection to family members, public and environment. Mathieu *et al.* reported the dose to family members and children of thyroid cancer patients, and in all cases the measured doses were lower than 0.5 mSv [6].

4.2 Hyperthyroid patients

Table 2 presents the effective doses to relatives of the hyperthyroid patients treated with iodine-131 and measured with TLD dosimeters. Patient's relatives wore the TLDs for a period of one week. The mean value of the effective dose for the relatives of the hyperthyroid patients was 0.87 mSv. The range of the effective dose varied from 0.12 mSv to 6.79 mSv. For three patient spouses, the measured value was 0 mSv, and the explanation was that the patients stayed in different rooms and most of the times were away from home. The spouse of patient number 18 received the highest dose (6.79 mSv). The explanation was that the woman did not follow the given recommendation. She stayed very close to her husband all the time after he received his treatment. According to BSS, this 69-year-old woman was still below the 15 mSv allowed dose constraint. Further, eleven family members received effective doses higher than 1 mSv but less than 3 mSv (allowed dose constraint).

Figure 2 presents the values of all the hyperthyroid patients together with the dose constraint of the BSS. All values, except one, are below 3 mSv.

The statistical analysis showed significant difference between the relatives of thyroid cancer patients and hyperthyroid patients groups (Mann-Whitney U-Test $p < 0.001$).

The results of this study show that doses to the relatives of hyperthyroid patients are higher than the doses to the relatives of thyroid cancer patients. This observation is due to the lower retention and the faster wash-out of the iodine-131 activity from the body of the thyroid carcinoma patients, in spite of the higher administered activity.

Furthermore, the thyroid cancer patients after the administration of the radioiodine were hospitalized for three days in an isolated room. Another reason is that thyroid cancer patients retain less iodine as a result of the minimal thyroid tissue left after surgery. On the other hand, hyperthyroid patients were administered with fewer doses but were treated on an outpatient basis.

Journal of Physics: Conference Series **317** (2011) 012012 doi:10.1088/1742-6596/317/1/012012

Table 2. Effective doses (EF) to family members of hyperthyroid patients.

No.	EF (mSv)	Sex	Age (y)	No.	EF (mSv)	Sex	Age (y)
1	1.25	m	63	16	1.90	m	55
2	0.00	m	40	17	0.33	f	52
3	0.00	m	29	18	6.79	f	69
4	0.78	f	44	19	0.21	m	56
5	0.23	m	45	20	1.23	m	66
6	0.52	m	41	21	0.38	f	33
7	0.23	m	52	22	0.40	f	44
8	0.12	m	39	23	0.21	m	57
9	1.2	f	44	24	0.00	m	40
10	1.14	m	58	25	0.51	f	43
11	0.52	m	80	26	1.25	f	57
12	0.54	m	40	27	1.48	f	50
13	0.17	m	32	28	1.32	f	42
14	0.16	m	56	29	1.70	f	47
15	0.48	m	70	30	1.17	f	60
				Mean	**0.87**		

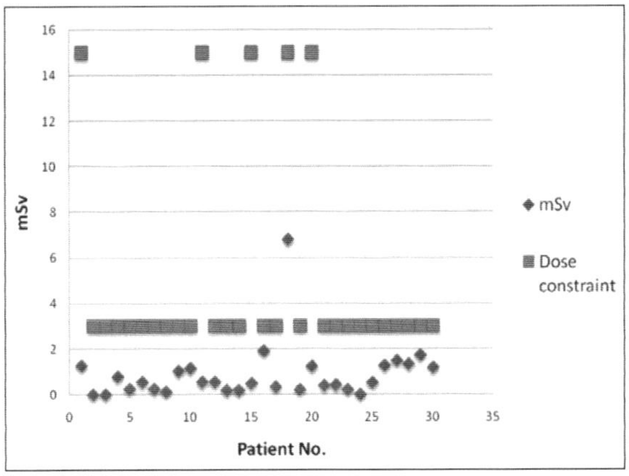

Figure 2. Effective doses to the relatives of hyperthyroid patients and the dose constraints of 3 mSv for adults and 15 mSv for older than 60 years of age, according to the BSS.

The oral and written instructions were the same, for both hyperthyroid and thyroid cancer patients. The restriction time is different (seven days for thyroid cancer patients and three days for hyperthyroid patients), and this information is usually given verbally by the physicians. Although the radiation doses to hyperthyroid family members are within the recommended limits, the values are higher in comparison with the doses of the family members of thyroid cancer patients. The higher doses to the relatives of the hyperthyroid patients are also due to the higher radioiodine retention of the thyroid gland and, thus, higher doses to the patients [6].

5. Conclusion

The radiation doses to family members of patients treated with radioiodine-131 for thyroid cancer were found to be well below the proposed dose limit of 1 mSv. Thyroid cancer patients should continue to be treated with the same way, so as patients after three days of hospitalization not to present any radiation hazard to their family members.

The effective doses of eleven family members of hyperthyroid patients were higher than 1 mSv, whereas one person received 6.79 mSv. It is necessary to formulate new instructions for hyperthyroid out-patients after treatment with radioiodine-131 to comply with the requirements of the revised ICRP limits.

This study has provided useful information on radiation protection and exposure to family members of patients with thyroid carcinoma and hyperthyroid, treated with radioiodine-131.

References

[1] International Commission of Radiological Protection 1991 *Recommendations of the International Commission on Radiological Protection* Ann. ICRP **21**(1-3)
[2] Council of the European Union 1996 *Guideline 96/29/EURATOM to determine basic directions for health protection of public and workers against radiation related dangers* Leafl. Eur. Commun. L159/1
[3] Monsieurs M *et al* 1998 Real-life burden to relatives of patients treated with iodine-131: a study in eight centers in Flanders (Belgium) *Eur. J. Nucl. Med.* **25** 1368-76
[4] Barrington S F *et al* 1999 Radiation exposure of the families of outpatients treated with radioiodine (iodine 131) for hyperthyroidism *Eur. J. Nucl. Med.* **26** 686-92
[5] Barrington S F *et al* 1995 Radiation exposures of patient's children at home post I-131 administration *Eur. J. Nucl. Med.* **22** 798
[6] Mathieu I *et al* 1997 Doses in family members after ^{131}I treatment *Lancet.* **350** 1074-5
[7] O'Doherty M J *et al* 1993 Radiation dose rates from adult patients receiving ^{131}I therapy for thyrotoxicosis *Nucl. Med. Commun.* **13** 160-8
[8] Cappelen T and Unhjem J F 2004 Patients with hyperthyroidism treated with radioactive iodine and exposures to family members *Eur. J. Nucl. Med. Mol. Imaging* **31** P828 S457
[9] Athanasopoulou E M *et al* 2007 Unexpected dose to the daughter of a patient treated with iodine-131 for hyperthyroidism *Hell. J. Nucl. Med.* Vol **10** 175-6
[10] Jacobsen A P, Plato P A and Toeroek D 1978 Contamination of the home environment by patients treated with iodine-131: initial result *Am. J. Public Health* **68** 225-30
[11] Buchan R C T and Brindle J M 1971 Radioiodine therapy to out-patient - the radiation hazard *Br. J.Radiol.* **44** 973-5

Re-186 and Sm-153 dosimetry based on scintigraphic imaging data in skeletal metastasis palliative treatment and Monte Carlo simulation

M Andreou, N Lagopati and M Lyra

Radiation Physics Unit, A′ Radiology Department, Kapodistrian University of Athens, 76 Vas. Sophias Ave, Athens, Greece

E-mail: mlyra@med.uoa.gr

Abstract. Optimum treatment planning of patients suffering from painful skeletal metastases requires accurate calculations concerning absorbed dose in metastatic lesions and critical organs, such as red marrow. Delivering high doses to tumor cells while limiting radiation dose to normal tissue, is the key for successful palliation treatment. The aim of this study is to compare the dosimetric calculations, obtained by Monte Carlo (MC) simulation and the MIRDOSE model, in therapeutic schemes of skeleton metastatic lesions, with Rhenium-186 (Sn) -HEDP and Samarium-153 -EDTMP. A bolus injection of 1295 MBq (35mCi) Re-186-HEDP was infused in 11 patients with multiple skeletal metastases. The administered dose for the 8 patients who received Sm-153 was 1 mCi /kg. Planar scintigraphic images for the two groups of patients were obtained, 24 h, 48 h and 72 h post injection, by an Elscint Apex SPX gamma camera. The images were processed, utilizing ROI quantitative methods, to determine residence times and radionuclide uptakes. Dosimetric calculations were performed using the patient specific scintigraphic data by the MIRDOSE3 code of MIRD. Also, MCNPX was employed, simulating the distribution of the radioisotope in the ROI and calculating the absorbed doses in the metastatic lesion, and in critical organs. Summarizing, there is a good agreement between the results, derived from the two pathways, the patient specific and the mathematical, with a deviation of less than 9% for planar scintigraphic data compared to MC, for both radiopharmaceuticals.

1. Introduction

Skeletal metastases, appears in the majority of patients with breast and prostate cancer and frequently in patients with lung, renal, thyroid cancer and multiple myeloma. Actually, two-thirds of patients with breast cancer develop metastatic bone disease. Delivering high doses to tumor cells while limiting radiation dose to normal tissue, is the key for successful palliation. Consequently, optimum treatment planning of patients suffering from painful skeletal metastases requires accurate calculations concerning absorbed dose in metastatic lesions and critical organs, such as red marrow.

The use of therapeutic radionuclides, which have the property of localizing in metastatic sites, has proved successful for alleviation of pain, especially for multiple sites for which the use of external beam irradiation is impractical. Systemic metabolic radiotherapy uses a variety of radionuclides conjugated to pyrophosphate analogues.

The aim of this study is to make a comparison between the dosimetric calculations, obtained by Monte Carlo (MC) simulation and the MIRDOSE model, in therapeutic schemes with Rhenium-186 (Sn) - HEDP (1,1 hydroxyethylidene diphosphonate), Re-186 – HEDP, and Samarium-153 -EDTMP (ethylenediamine tetramethylene phosphonate), Sm-153-EDTMP.

Samarium-153 - EDTMP is a 1:1 complex of the radioisotope Samarium-153 and a Tetraphosphonate (ethylenediamine-tetramethylene phosphonic acid). Samarium-153 - EDTMP concentrates in areas of high metabolic activity in skeletal tissue, where it is associated with the hydroxyapatite crystal [1]. It is injected into a vein and distributes throughout the body. It homes in on metastatic lesions. Once there, the radioisotope emits beta particles, which kill the nearby cancer cells. Pain begins to improve in the first week for most people and the effects can last several months [2]. Samarium-153 has a half life of 46.3 h, with maximum beta-particle energies of 810 keV (20%), 710 keV (49%), 640 keV (30%) and gamma photon emission of 103 keV (28%) which is suitable for imaging purposes during therapy. The short half-life of 46 h and its beta emissions allow the delivery of a high dose rate in regions of enhanced osteoblastic activity over a short period of time. This fact accounts for the reduced residual long-term activity in the bone marrow. Mean absorbed dose, a widely used parameter for prediction of the efficacy of a treatment can either overestimate or underestimate actual doses delivered in the organs/tissues of interest (TOIs), so it is far from being perfect.

^{186}Re (Sn) – 1,1 hydroxyethylidene diphosphonate, ^{186}Re – HEDP, is a radiopharmaceutical that combines selective localization in osteoblastic skeletal metastases with favorable radiation characteristics concerning pain palliation as well as dosimetric estimations and scintigraphic imaging. Unavoidable red marrow toxicity is limited to transient and reversible thrombocytopenia while leucopenia plays only a minor role [3]. The relatively short physical half-life combined with the beta emissions allow the delivery of relatively high dose rate for a short period of time in the areas of concentration.[4, 5]. ^{186}Re is a mainly beta-emitting radionuclide with a physical half-life of 89.3 h (3.78 d). Its main beta-emissions have maximum energies of 1.077 MeV (71%) and 0.939 MeV (22%) respectively [6-9].

2. Methods and materials

A bolus injection of 1295 MBq (35 mCi) Re-186–HEDP was infused in 11 patients with multiple skeletal metastases. It is proved to be a highly justified and efficient dose. The administered dose for the 8 patients, who received Sm-153, was 1 mCi/kg.

Planar scintigraphic images (anterior and posterior) for the two groups of patients, (Re-186 and Sm-153 patients), were obtained 24 h, 48 h and 72 h post injection, by an Elscint Apex SPX gamma camera. The images were processed utilizing ROI (regions of interest) quantitative methods to determine residence times and radionuclide uptakes by TOIs and by specific skeletal sites. Dosimetric calculations were performed using MIRDOSE3.

The analysis of these values lead to calculations of two important parameters: metastatic/normal bone absorbed dose ratio (M/N ratio) and bone/red marrow mean absorbed dose ratio (B/RM ratio). M/N ratio provides valuable information in assessing tumor-control probability, normal tissue toxicity and radiopharmaceuticals' qualification and superiority whereas B/RM ratio displays the red marrow toxicity induced by the radiopharmaceutical, a key issue for the success of the radiopharmaceuticals' therapeutic use. The bone ratios M/N and B/RM deduce the selective concentration of ^{186}Re – HEDP in metastatic lesions and conserving normal bone tissue.

2.1. Monte Carlo simulation

To prepare the input file for the MCNPX code (Monte Carlo N - Particle Extended Code), it is obligatory to define the geometry of the theoretical experiment. The metastatic lesion was considered to be a sphere of 1.5 cm radius, inside a cylindrical form simulating the bone. The spherical cells inside and outside the tumour, were defined from 0 up to 1 cm every 0.2 cm, from 1 up to 3 cm every 1 cm and from 3 up to 10 cm every 4 cm.

The radioactive source was considered as a point source, with the appropriate energy definition, according to the simulated radiopharmaceutical. It is identified with the center of the tumour. One million events were counted for both cases.

Thus, MCNPX (Monte Carlo N - Particle Code) code was employed, simulating the distribution of the radioisotope in the ROI and calculating the absorbed doses in metastatic lesion in critical organs.

3. Results

Gamma camera images of radionuclide distribution (whole-body scintigrams & SPECT) of patients, who were administered Re186 - HEDP and Sm153- EDTMP, were analyzed to measure activity in specifically selected normal and metastatic regions of interest [10]. Calculations based on the MIRD schema, gave values of absorbed dose per unit volume (voxel) for metastatic and normal bone tissue for the two studied radiopharmaceuticals.

The calculated values are presented in table 1. The absorbed dose of metastatic lesion over normal bone tissue per volume ratio, M/B ratio, represents the absorbed dose to a target volume compared with normal bone tissue and provides valuable information in assessing tumor control probability and normal tissue toxicity as well as the radiopharmaceutical's qualification and superiority. Bone tissue mean absorbed dose per red marrow absorbed dose ratio, B/RM ratio, is also presented in table 1 to display the radiopharmaceutical's toxicity on this critical organ during metastatic bone lesion therapy.

Table 1. Bone seeking radiopharmaceuticals' absorbed dose ratios: Metastatic/Normal bone bone tissue (M/B), Bone tissue/Red Marrow (B/RM).

	Absorbed Dose (mGy/MBq per volume unit)			
Radiopharmaceutical	Normal Bone	Metastatic lesion	M/B ratio	B/RM ratio
Re-186-HEDP	3.12	16.2	5.2	3.4
Sm-153-EDTMP	3.80	22.8	6.0	5.5

Table 1 compares the M/B and B/RM ratios for the 2 most commonly used radiopharmaceuticals in metastatic bone therapy. Sm-153-EDTMP provides an optimum combination of high M/B ratio (high therapeutic efficacy) as well as high B/RM ratio (low toxicity).

A deviation of about 7% between the results, obtained by the MIRDOSE and Monte Carlo results was found, in both cases (Rhenium and Samarium). Furthermore absorption linearity of radiation was observed, with a maximum distribution of the dose at the center of the tumour and gradual decrease towards the periphery. This proves that Sm-153-EDTMP is the radionuclide of choice.

4. Discussion

A method of estimating the optimum dosage of the administrated radiopharmaceutical in order to avoid myelotoxicity is the acquisition and quantification of whole body scintigraphic images using a smaller quantity of the radionuclide. By administering 740 MBq of ^{153}Sm-EDTMP anterior and posterior scintigraphic whole body images were obtained at 10 min and 5 h post-administration. Bone and bone marrow activity were calculated using the MIRD schema. The total activity delivering a limiting dose of 2 Gy at the bone marrow was determined. The total activity predicted was between 35% and 63% of the standard recommended value of 37 MBq/kg, which if administered, would have delivered bone marrow doses of 3.27-5.9 Gy resulting in myelotoxicity [11]

Toxicity side effects are also dose related. Dose distribution in individual bone marrow cavities, irradiated by a Sm-153 source and uniformly distributed on the bone surface, is different in reality from the uniform dose distribution, which is usually assumed. This difference affects the corresponding haematopoietic stem cell survival. In a study where Monte Carlo simulations were used to generate dose-volume histograms for a geometrical model of trabecular bone [12], the estimated cell survival for Sm-153 was markedly higher compared to a homogeneous dose distribution.

Therefore, the more realistic dose distribution models for individual bone marrow cavities can result in more accurate estimations of bone marrow toxicity.

However, with ^{186}Re-HEDP for palliative bone metastases therapy, significant pain relief (at 75% of patients) was observed. The results appear shortly after the 1st administration (1st week) and could last up to the 16th week. Multiple administrations are possible with minimum toxicity effects [3, 13-14].

Because ^{186}Re-HEDP delivers a substantial dose to bone marrow, bone marrow toxicity will be the dosage limiting factor. In most cases marrow toxicity is limited to temporary myelo-suppression confined mainly to thrombocytopenia while leucopenia plays only a minor role. In a previously published study of 60 patients with painful bone metastases from different tumor types, who were treated with 1406 MBq of ^{186}Re-HEDP, a WHO grade 1 - 2 hematologic toxicity was apparent with a decrease in the mean platelet (32%) and mean leukocyte (18%) counts at 3 and 4 weeks, respectively. In all patients, platelet and white blood cell counts returned to baseline levels within 8 weeks after administration of ^{186}Re-HEDP [15].

5. Conclusion

Summarizing, the mean absorbed dose, usually results in overestimations and underestimations of actual doses in TOIs. Dose distribution and time dose rate curves in TOIs are crucial parameters for the assessment of the relative biological effectiveness of either an ongoing treatment or of a treatment that is about to start. There is a good agreement between the results, derived from the two pathways, the real and the theoretical, with a deviation less than 9% for planar data comparing to MC, for both radiopharmaceuticals.

References

[1] Serafini A N 2001 *Q. J. Nucl. Med.* **45(1)** 91-9
[2] Bolch W E *et al* 1999*J. Nucl. Med.* **40(1)**
[3] Maxon H R *et al* 1991 *J. Nucl. Med.* **32** 1877-81
[4] Ketring A R 1987 *Nucl. Med. Biol.* **14** 223-32
[5] Maxon H R *et al* 1992 *Semin. Nucl. Med.* **22** 33–40
[6] De Klerk J M H *et al* 1996 *J. Nucl. Med.* **37** 38-41
[7] Maxon H R *et al J. Nucl. Med.* **32** 1877-81
[8] Sciutto R *et al* 2000 *J. Nucl. Med.* **41** 647-54
[9] Lyra M *et al* 1999 *Eur. J. Nucl. Med.* **26** 191
[10] Lyra M, Limouris GS, Frantzis A, Phinou P, Emmanouelides J, Voliotopoulos B , Stavraka A, and Vlahos L 1999 *Eur. J. Nucl. Med.* **26(9)** 1191
[11] Cameron P J, Klemp P F, Martindale A and Turner H 1999 *Nucl. Med. Commun.* **20(7)** 609-15
[12] Kvinnsland Y, Skretting A and Bruland O S 2001 *Phys. Med. Biol.* **46(4)** 1149-61
[13] Fitzpatrick P J and Rider W D 1976 *Int. J. Radiat. Oncol. Biol. Phys.* **1** 197-207
[14] Gilbert H A *et al* 1977 *Am. J. Roentgenol.* **129** 1095-6
[15] Sciutto R *et al* 2000 *J. Nucl. Med.* **41** 647-54

Journal of Physics: Conference Series **317** (2011) 012014 doi:10.1088/1742-6596/317/1/012014

Pitfalls in classical nuclear medicine: myocardial perfusion imaging

C Fragkaki[1] and Ch Giannopoulou[2]

[1] Nuclear Medicine Department, EUROMEDICA / Ambelokipoi Branch, Athens, Greece
[2] Nuclear Medicine Department, "Evangelismos" General Hospital, Athens, Greece

E-mail: christina.fragkaki@gmail.com

Abstract. Scintigraphic imaging is a complex functional procedure subject to a variety of artefacts and pitfalls that may limit its clinical and diagnostic accuracy. It is important to be aware of and to recognize them when present and to eliminate them whenever possible. Pitfalls may occur at any stage of the imaging procedure and can be related with the γ-camera or other equipment, personnel handling, patient preparation, image processing or the procedure itself. Often, potential causes of artefacts and pitfalls may overlap. In this short review, special interest will be given to cardiac scintigraphic imaging. Most common causes of artefact in myocardial perfusion imaging are soft tissue attenuation as well as motion and gating errors. Additionally, clinical problems like cardiac abnormalities may cause interpretation pitfalls and nuclear medicine physicians should be familiar with these in order to ensure the correct evaluation of the study. Artefacts or suboptimal image quality can also result from infiltrated injections, misalignment in patient positioning, power instability or interruption, flood field non-uniformities, cracked crystal and several other technical reasons.

1. Introduction

Myocardial perfusion imaging (MPI) has established its utility and its clinical importance in the diagnoses of several medical conditions throughout the years and today is one of the most common examinations requested at a nuclear medicine department. It is important though to be aware that a variety of artefacts and pitfalls that might occur at any stage during imaging and nuclear medicine physicians should be familiar with these to ensure a correct evaluation of the study. In this short review most of these artefacts and pitfalls will be addressed.

2. Clinical artefacts and pitfalls

2.1. Soft tissue attenuation artefact

Attenuation is the decrease in intensity of a photon signal along its path to the detector. During nuclear cardiac imaging, non-uniform attenuation occurs as photons pass through tissues of varying densities. Artefact extent varies with the location of soft tissue, overall patient body size, and depth of target organ. In particular:

- Breast usually causes defects in the anterior wall but if breasts are pendulous, lateral defects may also occur,
- Obesity is usually associated with lateral wall fixed defects,
- Diaphragm causes inferior wall defects,
- Prone SPECT can reduce diaphragmatic attenuation but may lead to small anterior and anteroseptal defects.

2.2. Overlying visceral activity

SPECT reconstruction leads to apparent increased count densities in areas underlying visceral activity, since polar maps are normalized to areas with highest activity, bull-eye displays can be affected, as shown in figure 1.

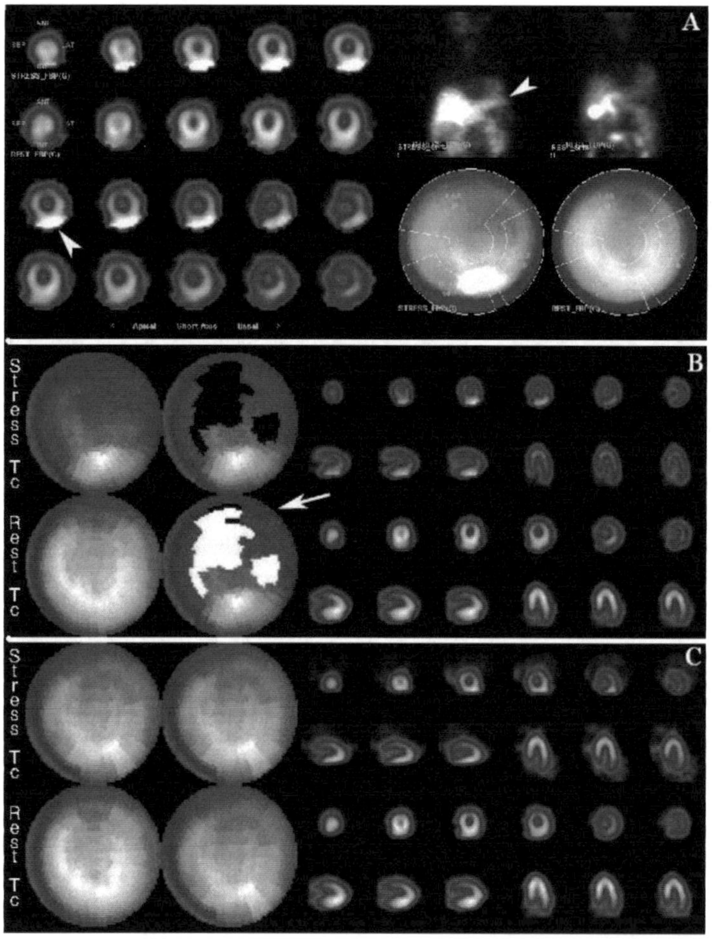

Figure 1. (A) An example of a patient with intensevisceral activity (arrowheads). (B) Image processing which includes this activity results in an artefactual reversible defect (arrow) on perfusion images. (C) The artefactual defect disappears when the study is reprocessed correctly showing a normal study.

2.3. Myocardial hot spots

Myocardial hot spots are due to increased myocardial thickness near the anterior and posterior papillary muscles, exacerbated by left ventricular hypertrophy. Bull-eye reconstruction can therefore be affected.

2.4. Artefacts due to patient's anatomy
Anatomical variants or other medical conditions can affect cardiac imaging:

- Apical variants (apex displaced laterally or rarely septally) make selection of the apex difficult and may lead to bull-eye reconstruction errors,
- LBBB causes reversible septal defects. ECGs should be viewed for conduction abnormalities before the stress test and if present always do adenosine or dipyridamole stress test,
- Left ventricular hypertrophy (LVH) can simulate lateral infarction,
- Dextrorotation can cause lateral wall defect and Levorotation a septal wall defect. If the condition is known beforehand there should be a 360° SPECT performed,
- In cases of dextrocardia the 180° SPECT acquisition should range from 135° to +45° [1].

2.5. Motion artefacts
Patient motion during myocardial perfusion SPECT can produce artefactual perfusion defects. Motion artefacts appear to be a greater problem with axial than lateral movement and are accentuated if movement occurs in the middle of rotation. Horizontal motion typically causes septal and lateral artefacts and vertical motion may cause anterior and inferior wall artefacts [2].

2.6. Upward creep artefact
During exercise, the heart is displaced interiorly against the diaphragm as the thorax expands. During recovery, the heart "creeps" back up in the thorax leading to reversible defects in the inferoseptal wall. This artefact can be avoided if we wait for about 15 min before starting the stress imaging [3].

3. Technical artefacts and pitfalls

3.1. Radiolabeling errors
Inadequate binding of the technetium to the cold ingredient will result in free ^{99m}Tc in the stomach causing an artefact and non-imaging of the heart.

3.2. Infiltrated injection
In case of subcutaneous radiopharmaceutical infiltration, less radiopharmaceutical is taken up by the myocardium resulting in lowered counting statistics and a poor-quality study. Should the infiltrated injection occur during the second phase of a same–day Rest–Stress study, the stress scan will be predominated by activity from the first injection (for the rest imaging) and ischemia induced during the stress test may be masked [1].

3.3. Hot spots near the heart
Increased uptake adjacent to the heart could conceivably compromise evaluation of perfusion in adjacent LV. An example is seen in figure 2.

3.4. Metal and other attenuators
Before imaging, metal and other potential attenuators must be removed from the patient if they will project into the imaging field of view and potentially interfere with the study.

3.5. Contamination
Any contamination with the radiopharmaceutical on the patient or the equipment can project into the imaging field of view and interfere with the study.

3.6. Processing errors

- Incorrect parameter setting or incorrect axis alignment can lead to wrongful evaluation of the study,

Journal of Physics: Conference Series **317** (2011) 012014 doi:10.1088/1742-6596/317/1/012014

- Filters that are too coarse may result in false positive studies and filters that are too smooth may result in false negative studies [4],
- A large number of beat rejections has a great impact on the gated data set and the QC of the gating data should always be checked before reporting a study.

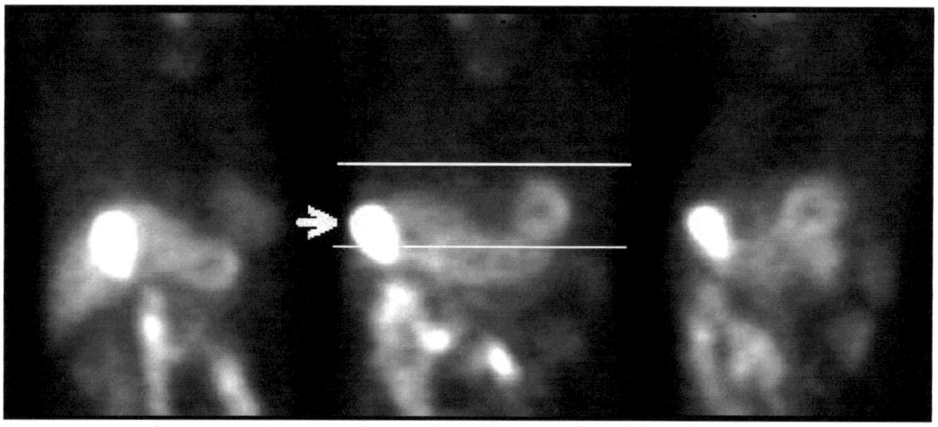

Figure 2. A very "hot" and upward positioned gall bladder (arrow) projecting into the imaging field of view can compromise the imaging of the heart.

3.7. Center of rotation artefact
If the center of rotation is properly calibrated, a radioactive point source placed in the center of the camera orbit should project to the center of the computer matrix. If the calibration is incorrect a posteroapical (rightward +) or an anteroapical (leftward –) defect will result [5].

3.8. γ-ray scatter artefact
When imaging using ^{99m}Tc, incident photons contributing to the final image may include photons scattered by as much as 45° to 60°, so care should be taken to clear away any external sources [6]. Unfortunately, several other technical and acquisition issues can arise (like flood field non-uniformities, power interruption to the camera, unexpected photopeak shift, cracked crystal etc) and is crucial to be ready to recognize them as soon as possible.

4. Conclusion
As with all medical procedures, rashness and lack of attention to details may lead to pitfalls that could be easily avoidable. Should artefacts occur they should be recognized and corrected on time. Using standard protocols, being thorough with the quality control of the equipment and the standard operating procedures in a department can help to prevent most of these artefacts and pitfalls.

References
[1] Burrell S and MacDonald A 2006 *J. Nucl. Med. Technol.* **34** 193-211
[2] Cooper J A, Neumann P H and McCandless B K 1992 *J. Nucl. Med.* **33** 1566-71
[3] Matsumoto N, Berman D S, Kavanagh P B, Gerlach J, Hayes S W, Lewin H C, Friedman J D and Germano G 2001 *J. Nucl. Med.* **42** 687-94
[4] Wheat J M and Currie G M 2007 *Internet J. Cardiovas.* **4(1)**
[5] Mettler F A and Guiberteau M J 1998 *Essentials of Nuclear Medicine Imaging* 4th edition (Philadelphia: W.B. Saunders) pp 129–189
[6] Currie G M and Wheat J M 2008 *Internet J. Med. Technol.* **4(2)**

Pitfalls in PET/CT imaging

Ph Rondogianni[1], N Papathanasiou[2] and Ch Giannopoulou[1]

[1] "Evangelismos" General Hospital, Athens, Greece
[2] "Metropolitan" General Hospital, Piraeus, Greece

E-mail: phrontog@yahoo.gr

Abstract. PET with 2-[fluorine 18] fluoro-2-deoxy-d-glucose (FDG), has been a clinical tool for the evaluation of various cancers providing valuable metabolic information clinically helpful in the diagnosis, initial staging, therapy monitoring and restaging. However, FDG is not specific for neoplastic processes. Unless anatomic correlation is available to delineate normal structures, pathologic sites of FDG accumulation can easily be confused with normal physiological uptake, leading to false-positive or false-negative findings. Co-registration of PET scans (functional and morphologic information) with computed tomographic (CT) scans (anatomic information) using a combined PET-CT scanner improves the overall sensitivity and specificity of information provided by PET or CT alone. In this paper, we discuss the probable causes of false negative images and pitfalls due to technical reasons, inflammatory processes or benign lesions as well as the utility of PET-CT in differentiating malignant from inflammatory and benign processes, since in some cases such differentiation cannot be made, with certainty, using FDG PET alone.

1. Introduction

Since the middle of 1995 Positron Emission Tomography (PET) with 2-[fluorine 18] fluoro-2-deoxy-d-glucose (FDG), has been a clinical tool for the evaluation of various cancers providing valuable metabolic information clinically helpful in the diagnosis, initial staging, therapy monitoring and restaging. FDG uptake and accumulation in neoplastic tissues is due to the fact that the vast majority of malignant cancer phenotypes exhibit an increased glycolytic rate (Warburg effect) [1]. This is because malignant cells have substantially enhanced glucose transporters on their surface (primarily GLUT-1), and express high levels of hexokinase and low levels of glucose-6 phosphatase, as a result of which FDG accumulates in these cells. Well recognized advantages of FDG-PET over anatomic based cross-sectional imaging modalities such as computed tomography (CT) and conventional (T1- and T2-weighted) magnetic resonance imaging is the high sensitivity due to its high lesion to background contrast and its ability to detect metabolic changes that may occur well before gross structural or contrast enhancement changes become visible [1-2].

FDG is not specific for neoplastic processes; it accumulates physiologically in various normal organs, including the brain, muscles, salivary glands, myocardium, gastrointestinal tract, urinary tract, brown adipose tissue, thyroid gland, and gonadal tissues. It is also taken up by different inflammatory as well as benign lesions. Unless anatomic correlation is available to delineate normal structures, pathologic sites of FDG accumulation can easily be confused with normal physiological uptake, leading to false-positive or false-negative findings. This is an important shortcoming in the determination of active disease sites, particularly for small lesions and lesions located near sites of physiological uptake. Coregistration of PET scans (functional and morphologic information) with computed tomographic (CT) scans (anatomic information) using a combined PET-CT scanner improves the overall sensitivity and specificity of information provided by PET or CT alone [3, 4]. The unique advantage of PET-CT fusion imaging is the ability to correlate findings from two complementary imaging modalities in a comprehensive examination. Hence, PET-CT provides more precise anatomic definition for both the physiologic and pathologic uptake seen with FDG PET. Conversely, equivocal CT findings can be better evaluated with the help of the additional functional information provided by FDG PET.

In this paper, we discuss the probable causes of false negative images and pitfalls due to technical reasons, inflammatory processes or benign lesions as well as the utility of PET-CT in differentiating malignant from inflammatory and benign processes, since in some cases such differentiation cannot be made, with certainty, using FDG PET alone.

2. Discussion

2.1. Normal physiological FDG distribution

As a general rule, physiological FDG uptake follows the distribution pattern of glucose-metabolizing cells and organs. FDG shows avid physiological accumulation in the brain, because brain satisfies its high energy demand almost entirely with glucose, regardless of the substrate provided. This leads to a suboptimal contrast resolution for brain PET, which is inadequate for the detection of small lesions and low-grade tumors.

In the head-neck region, there are various structures showing normal increased FDG uptake. The salivary glands (parotids, submandibular and sublingual glands) may show moderate or intense activity. Lymphoid tissue in the Waldeyer ring (palatine and lingual tonsils) often shows increased activity, especially in kids, with peak at 6-7 years of age. Talking or crying during the uptake phase may induce FDG accumulation in the tongue, vocal cords and arytenoid muscles. Chewing gum or jaw movements may cause uptake within muscles of mastication. A symmetrical or almost symmetrical uptake pattern in head-neck structures is usually within normality [5].

Essentially, muscular FDG uptake is low when the muscles are at rest; however intense muscular activity prior to the study or during the uptake phase may induce increased uptake. A pattern of diffuse muscular FDG uptake may be encountered as a result of hyperglycemia. Increased blood glucose has a deleterious effect on PET imaging for 2 reasons: (a) excessive “cold” glucose competes with “hot” FDG and may saturate glucose transporters of malignant cells, (b) hyperglycemia triggers high levels of insulin, which shunts FDG into skeletal muscles, hence compromising the quality and sensitivity of the scan. A heavy meal before the scan may induce the same response: increase insulin levels and FDG uptake in the skeletal muscles [6]. To prevent such an effect, specific attendance should be given in fasting instructions and a priori adequate glucose control in diabetics.

Brown fat FDG uptake is a frequent pitfall of PET-CT imaging. Brown fat is a thermogenic tissue, with activation and increased consumption of glucose and, consequently, FDG as a response to acute cold. In such cases, there is a symmetrical pattern in cervical, supraclavicular, axillary, paraspinal and perirenal regions, corresponding to areas of fat attenuation on CT [7].

Myocardial FDG activity is quite variable ranging from non-discernible to intense, throughout the left ventricle. Uptake depends on fasting state, substrate availability and insulin levels: in a fasted patient with low level of insulin, myocardium uses fatty acids, not glucose, as primary energy source. FDG shows normal homogeneous uptake in the liver. In the gastrointestinal tract, physiological uptake is highly variable in distribution and intensity. Usually, diffuse colonic uptake is normal, whereas focal uptake should be further investigated with colonoscopy to exclude malignancy [8]. Metformin medication is associated with a diffuse intense bowel uptake pattern.

FDG is excreted in the urine, without being reabsorbed in renal tubules. Thus, any part of the urinary tract may show increased FDG activity. Dilated ureters, bladder diverticulae, surgical urinary diversions (ileal conduits) or any other cause of retained activity may be confused with a lesion. Intense urinary bladder activity may obscure a true lesion in adjacent nodes or organs including uterus, ovaries and prostate. Good hydration and voiding before the scan can help minimize this effect [8].

Finally, FDG uptake in the genital tract should be interpreted in the context of menstrual status. In premenopausal women, normal focal FDG uptake is often identified in the ovaries during the midcycle periovulatory phase. Normal endometrial uptake may occur in the menstrual flow phase or less frequently the ovulating phase.

2.2. False negative findings

Despite the high sensitivity of FDG PET/CT, there are false negative results due to different causes like tumor biology, small tumor size under the system resolution, improper patient preparation or recent chemotherapy.

The vast majority of malignant tumors are FDG avid. Aggressive tumors tend to have higher levels of FDG uptake while less aggressive tumors tend to have lower levels of FDG uptake [9-10]. This dimension of diagnostic information provided by FDG-PET is very important because it can be used to improve determination of disease prognosis and treatment planning [11].Variables that influence FDG uptake are tumor Ki67 index, grade of differentiation, low malignant tumor cell density within the tumor mass or presence of large mucinous components, type of glucose transporters on the cell membrane and, levels and activity of glucose-6-phosphatase enzyme in the cell. Neoplasms that are frequently not hypermetabolic and thus not FDG avid, leading to false-negative results, include certain renal cell cancers and lymphomas, hepatocellular carcinomas, cholangiocarcinomas, neuroendocrine tumors, colonic mucinous adenoacarcinomas, prostate carcinomas, and carcinoid tumors.

Five functional mammalian facilitated hexose transporters (GLUTs) have been characterized by molecular cloning. By functional expression in heterologous systems, their specificity and affinity for different hexoses have been defined. There are three high-affinity transporters (GLUT-1, GLUT-3 and GLUT-4) and one low-affinity transporter (GLUT-2). GLUT-5 is primarily a fructose carrier. Because their Michaelis constants (Km) are below the normal blood glucose concentration, the high-affinity transporters function at rates close to maximal velocity. Thus, their level of cell surface expression greatly influences the rate of glucose uptake into the cells. In contrast, the rate of glucose uptake by GLUT-2 (Km = 17 mM) increases in parallel with the rise in blood glucose over the physiological concentration range. High-affinity transporters are found in almost every tissue, but their expression is higher in cells with high glycolytic activity. Glut-2, however, is found in tissues carrying large glucose fluxes, such as intestine, kidney, and liver. As an adaptive response to variations in metabolic conditions, the expression of these transporters is regulated by glucose and different hormones. Thus, because of their specific characteristics and regulated expression, the facilitated glucose transporters control fundamental aspects of glucose homeostasis [12].

Malignant cells have generally high-affinity transporters on the cell membrane. There are several tumors, however, like Hepatocellular Carcinoma (HCC) where GLUT 2 and GLUT5 transporters predominate. As a result, these tumors have low FDG uptake. Another cause of minimal or absent FDG uptake by HCC is the high level of glucose-6-phosphatase in liver tissue leads to the release of FDG-6-phosphate. Thus, 18F-FDG PET has an average false-negative rate of 40% to 50% for the detection of HCC [13], and data on the role of PET/CT in the detection of HCC metastasis are limited [14-15]. Nevertheless, it has been shown that the overall survival of patients with HCC but with negative or partially positive 18F-FDG findings is lower than that of patients with positive 18F-FDG findings.

In prostate cancer, the absence of FDG uptake is connected to a low Gleason score and to levels of Prostate Specific Antigen (PSA), and reflects less aggressive tumor behaviour [11]. Lymphomas are generally FDG avid tumors and consist the main indication for the prescription of an FDG PET/CT examination. There are, however, some indolent lymphomas, especially, the malt extranodal and the marginal zone splenic subtypes with low or even absent uptake. This is important to know as patients having such a histological lymphoma subtype can not be surveyed by FDG PET/CT. In aggressive lymphomas, however, FDG uptake may serve as a marker of disease aggressiveness. Maroun *et al* [16] have evaluated the association of the SUVmax of the 'hottest focus' with overall survival (OS) and failure-free survival (FFS) in patients with aggressive mantle lymphoma. They have shown that both

Journal of Physics: Conference Series **317** (2011) 012015 doi:10.1088/1742-6596/317/1/012015

the OS and FFS for patients with SUVmax greater than 5 were significantly decreased ($p < 0.01$ and < 0.001, respectively) as compared with the patients with SUV ≤ 5.

18F-FDG PET might be of value for neuroendocrine tumors with a high proliferation index, whereas the diagnostic sensitivity seems to be low for tumors with a low proliferation index, slow growth rate, and low glucose consumption [17].

FDG PET is at least as sensitive as bone scintigraphy in detecting the majority of skeletal metastases. It has higher sensitivity than 99mTc MDP scintigraphy in lytic metastases - especially of lung cancer - but may be less sensitive in predominantly sclerotic lesions [18].

Small tumor size is another cause of a false negative result for two reasons: firstly, in the case of hypermetabolic but small lesions adjacent to normal structures with high FDG uptake (brain, heart, kidneys) and, secondly, because of the partial volume effect**.** Partial volume effect provokes false negative images of tumors that are smaller than two times the spatial resolution of the scanner, especially in a milieu of low FDG uptake (i.e. lung) [19].

Chemotherapeutic agents may reduce FDG uptake by altering the glucose metabolism of tumor cells. It has been shown that there is 40% decrease in tumor cell hexokinase activity in tumors resected from patients who had undergone recent chemotherapy [20]. For this reason, it is important not to perform a PET/CT in less than ten days after chemotherapy. Interim studies must be carried out near the next therapeutic schema, if possible just the day before.

2.3. False positive findings

Despite the high sensitivity of FDG PET in malignancies, one of the inherent pitfalls of metabolic imaging has been the relatively low specificity with false-positive results due to increased FDG uptake in normal organs, inflammatory conditions, and benign processes, findings that can complicate image assessment.

FDG is not a cancer specific radiopharmaceutical. It accumulates physiologically in various normal organs, including the brain, heart, gastrointestinal tract, urinary tract, brown fat, and gonadal tissues, as well as in benign conditions such as infectious and inflammatory lesions and several benign tumors. False-positive FDG uptake can occur with PET-CT due to granulomatous disease, abscess, post-surgical changes, post-radiotherapy, foreign body reaction, inflammation (e.g., in diverticulitis, gastritis, or arteriosclerosis) or technical artefacts.

2.4. Technical artefacts

The artefacts most commonly seen on PET/CT images are due to attenuation correction, respiratory motion, and truncation. These artefacts occur because the CT scan is used instead of a PET transmission scan for attenuation correction of the PET data.

Attenuation correction refers to manipulation of the PET data to produce an image that takes into account the greater attenuation of photons that arise from deeper structures in the body or pass through highly attenuating structures such as bone on their route to the detector. It is generally achieved by applying an attenuation map to the PET data, derived from the CT image (after correction for the different photon energies used in both modalities). Attenuation correction artefacts can occur where there are highly attenuating objects in the path of the CT beam, such as dental fillings, hip prostheses, metallic stents, chemotherapy ports and high-density drainage tubes, and contrast-enhanced vessels. The CT-derived attenuation map corrects (or overcorrects) photopenic areas adjacent to high-attenuation structures at CT and makes them appear hypermetabolic on the attenuation-corrected PET images. Non-attenuated PET images, which do not manifest this error, can be used in these cases to aid the interpretation of these artefacts [21].

Respiratory motion during scanning causes the most prevalent artefact in PET-CT imaging. Because of the long acquisition time of a PET scan, it is acquired while the patient is freely breathing. The final image is hence an average of many breathing cycles. On the other hand, a CT scan is usually acquired during a specific stage of the breathing cycle. This difference in respiratory motion between PET scans and CT scans can cause misregistration on PET-CT images. This affects structures close to the diaphragm (i.e. lung nodules). In clinical practice, it most often causes difficulties in differentiating activity in the liver and adjacent structure [22].

Truncation artefacts are frequently seen in large patients or patients scanned with arms down, such as in the case of melanoma and head and neck indications. When a patient extends beyond the CT field of view, the extended part of the anatomy is truncated and consequently is not represented in the reconstructed CT image. This, in turn, results in no attenuation correction values for the corresponding region in the PET emission data, hence introducing a bias on the PET attenuation-corrected images that underestimates the standardized uptake values in these regions. Truncation also produces streaking artefacts at the edge of the CT image, resulting in an overestimation of the attenuation coefficients used to correct the PET data. This increase in attenuation coefficients creates a rim of high activity at the truncation edge, potentially resulting in misinterpretation of the PET scan [22 23].

2.5. Infection-Inflammation

Monocytes, macrophages and lymphocytes have Glut3 transporters in intracellular vesicles, in normal state. During their activation, these transporters are translocated fused to the cell membrane. As a result, high FDG uptake is observed in a number of infectious and inflammatory conditions as infected or inflamed sites [24].

Active granulomatous processes (tuberculosis, sarcoidosis), active fibrotic lesions and other infectious conditions (abscesses, pneumonia, sinusitis, gastritis, esophagitis, diverticulitis, appendicitis) are common causes of false positive images. Inflammatory changes from recent surgery may be FDG avid for a time longer than eight weeks while post radiation inflammation may last more than six months [25].

Degenerative or inflammatory joint disease can also give rise to elevated FDG uptake. Osteophytes can demonstrate FDG uptake, which depends on the degree of metabolic activity. Although osteophytes can occur at any level of the vertebral column, those that are located anteriorly may be confused with the paravertebral lymph nodes unless the CT images are evaluated on bone window settings

Healing bone is associated with elevated FDG uptake. Although the uptake in healing fractures is typically modest, it can be indistinguishable from small metastases. The origin of such uptake is unclear. Hematoma formation and the granulomatous tissue associated with resorption of the hematoma could account for the early phase of FDG uptake but the uptake observed weeks into the healing phase suggest that the procallus itself is associated with elevated glycolytic metabolism [26].

2.6. Benign tumors

Several benign tumors may also demonstrate moderate or even high FDG uptake. Such tumors are follicular thyroid adenomas, adrenal adenomas, uterine fibromas, Schwanomas, adenomatous polyps etc [27]. In such cases, the distinction between benign and malignant lesion may be impossible if CT image is not typical.

3. Conclusions

PET-CT using FDG is a very useful modality for cancer imaging. Combined PET-CT imaging permits easier recognition of normal variants and better localization and characterization of miscellaneous findings. Although the addition of anatomic CT information does aid in the interpretation of PET, artefacts can be introduced as a result of the dual-modality nature of the examination. To optimise interpretation, PET-CT images should be acquired after proper patient preparation and right quality control system, reviewed with the help of non-attenuation-corrected images, using all other available imaging and knowledge of the patient's history and system technical parameters.

REFERENCES

[1] Miles K A and Williams R E 2008 *Cancer Imaging* **8** 81-86
[2] Gillies R J, Robey I and Gatenby R A 2008 *J. Nucl. Med.* **49** 24-42S

[3] Goerres G W, von Schulthess G K and Hany T H 2002 *Am. J .Roentgenol.* **179** 1337-43
[4] Kluetz P G *et al* 2000 *Clin. Positron Imaging* **3** 223–230
[5] Shammas A *et al* 2009 *RadioGraphics* **29** 1467-86
[6] Cohade C 2010 *Semin. Nucl. Med.* **40** 283-293
[7] Virtanen K, Lidell M and Orava J 2009 *N. Engl. J. Med.* **360** 1518-25
[8] Liu Y, Ghesani N and Zuckier L 2010 *Semin. Nucl. Med.* **40** 294-315
[9] Jadvar H, Alavi A and Gambhir S S 2009 *J. Nucl. Med.* **50** 1820-7
[10] Basu S, Kumar R, Mavi A and Alavi A 2009 PET Clin. **38** 1–9
[11] Kwee T *et al* 2011 *Eur. J. Nucl. Med. Mol. Imaging* **38**(6) 1158-70
[12] Thorens B and Mueckler M 2010 *Am. J. Physiol. Endocrinol. Metab.* **298** E141–45
[13] Sugiyama M *et al* 2004 *J. Gastroenterol.* **39** 961-68
[14] Bohm B *et al* 2004 *J. Cancer Res. Clin. Oncol.* **130** 266–72
[15] Park J W *et al* 2008 *J. Nucl. Med.* **49** 1912-21
[16] Karam M, Ata A and Irish K 2009 *Nucl. Med. Commun.* **30** 770–78
[17] Binderap T *et al* 2010 *J. Nucl. Med.* **51** 704–12
[18] Cook G J *et al* 1998 *J. Clin. Oncol.* **16** 3375-9
[19] Soret M, Bacharach S and Buvat I 2007 *J. Nucl. Med.* **48** 932–45
[20] Akhurst T *et al* 2005 *J. Clin. Oncol.* **23** 8713-16
[21] McDermott S and Skehan S 2010 Abdom. Imaging **35** 55–69
[22] Sureshbabu W and Mawlawi O 2005 *J. Nucl. Med. Technol.* **33** 156–61.
[23] Gollub M J *et al* 2007 *J. Nucl. Med.* **48** 1583–91
[24] Paik J Y, Lee K H and Choe Y S 2004 *C. J. Nucl. Med.* **45** 124-28
[25] Vos F J *et al* 2006 *Q. J. Nucl. Med. Mol. Imaging* **50** 121-30
[26] Yiyan L 2009 *Ann. Nucl. Med.* **23** 701–708
[27] Kostakoglu L *et al* 2004 *RadioGraphics* **24** 1411–31

Journal of Physics: Conference Series **317** (2011) 012016 doi:10.1088/1742-6596/317/1/012016

Monte Carlo simulations for therapy imaging

M Ljungberg

Department of Medical Radiation Physics, Clinical Sciences, Lund, Lund University, Lund, Sweden

E-mail: michael.ljungberg@med.lu.se

Abstract. In therapy imaging there is a need to optimize imaging protocols and different correction methods for e.g. photon attenuation, contribution from scatter and other physics-related effects that limits the possibility to achieve accurate quantitative results. Physical phantom studies are important but lack flexibility and are often difficult to make patient realistic. A complement is to simulate the study using a Monte Carlo- based model of the imaging process including accurate description of the interaction processes and a realistic model of the patient and the imaging device. With the Monte Carlo method it is also possible to simulate and calculate parameters that otherwise would be impossible to measure from a physical experiment. An example of such is the simulation of a scatter and an attenuation-free situation. Results from such simulations can be very important in a validation procedure serving as reference data for which results obtained by different correction methods can be compared to. The Monte Carlo method has also been used to calculate absorbed doses from reference phantom or patient-specific phantom obtained from a CT study. This paper give an overview of the method and show examples of how Monte Carlo simulation can be used to evaluate therapy imaging with different radionuclides but also how simulations can be used for calculation of necessary parameters for correction of scatter, attenuation and collimator response.

1. Introduction

Selection of radionuclides for radionuclide therapy is, of course, optimized for a successful treatment and as a consequence these radionuclides are not always suitable for imaging with scintillation cameras. Limit factors can be multiple high-energy photon emissions or low emission abundances which can leading to noise problems. Nevertheless, imaging of on-going treatment can provide important information about the outcome of the treatment and also provide a verification that the predicted absorbed dose actually is delivered to the target volume and that normal organs/tissues, that are regarded as risk-organs, receive an absorbed dose that is acceptable and below levels for side-effects. In order to develop a proper method for quantification of activity it is important that the underlying problems are fully understood. A very useful tool to gain such knowledge is the Monte Carlo method. From Monte Carlo simulations, important parameters and imaging characteristics can be quantified and visualised and used to both develop and verify compensation methods. In this paper, we will discuss the application of Monte Carlo for the radionuclides ^{177}Lu, ^{131}I and ^{90}Y that are commonly used for different types of radionuclide therapies.

2. The Monte Carlo method

The Monte Carlo method is a statistical method that from random numbers perform simulation of stochastic problems using computers [9]. In medical imaging, the transport of particles (i.e. photon and electrons) can be simulated by a model of the system and the physical processes by sampling of probability density functions (*pdf's*). When comparing to experimental studies, Monte Carlo simulations have several advantages. Firstly, it is easy to modify the system and thus investigate the corresponding effect on the characteristics of the system. Secondly, it is possible to investigate parameters that are virtually impossible to measure. Examples of such parameters are the contribution to an image from photons scattered in the patient and the fraction of photons penetrating the collimator septa. Hence, a Monte Carlo program can provide information when developing correction methods since the physical processed generally in detail are known and components in, for example, an image can be separated allowing for studies of the relative importance of different effects.

There are several public-domain Monte Carlo programs available that can be used for Monte Carlo simulation of nuclear medicine imaging. The Gate program is a Monte Carlo program based on the Geant4 simulation toolkit and has been developed for tomographic applications [7]. It is based on a script-language and has a powerful geometry package making it very suitable for tomographic camera design for both SPECT and PET. EGS4 [12] and MCNPX [1] are programs mainly developed for high-energy physics but have capabilities of simulating imaging systems. At Lund University, we have developed the SIMIND code which is a Monte Carlo program dedicated for simulation of scintillation detectors, scintillation cameras and SPECT [8]. This program has been used in the following examples.

3. The scintillation camera

A standard scintillation camera consist of a NaI(Tl) crystal, a collimator of lead (mostly a parallel hole design) and a large set of photo-multiplication tubes (PMT) connected to position-recording electronics. When photons interact in the crystal, excitations occur with scintillation light emission. This light is guided to and collected by the PMT's to form an energy signal. The position of an interaction is determined by the energy signals from the different PMT's weighted by their spatial position. If the sum of the energy signals passes the energy discriminator the x,y position will be digitized and stored in an image matrix.

3.1. Planar imaging

A scintillation camera is a 2D imaging modality where images are created from photon travelling along rays (projections). This has some serious implications on the image mainly because no depth-information can be resolved. Contribution from overlapping and under-lapping activity can therefore mask small volume-of-interests making detection of small lesions impossible. The spatial resolution get worse at larger distances due to the design of the collimator which leads to loss in image contrast for sources distant to the camera surface. Furthermore, proper attenuation correction can be a problem.

3.2. SPECT

To overcome these problems, tomographic methods have been developed. The principle for SPECT (Single-Photon Emission Computed Tomography) is to measure planar images in different projections and from these projections reconstruct 3D tomographic images. Most methods used today in the clinic are based on the iterative MLEM/OSEM algorithms where projections of an initial guess of the activity distribution are determined iteratively from a relevant model of the imaging process in a forward projector step. Error projections are then obtained from the quota of the measured projections and the calculated projections. These error projections are back-projected to form an error image used to update the initial guess. This procedure continues until convergence has been reached. The differences between the MLEM and the OSEM algorithms are mostly related to when and how many stages, during a particular iteration, the estimate of the image should be updated since the error image in an OSEM method is calculated only from a subset of projection angles.

3.3. Photon attenuation
Measured projections are affected by non-homogeneous photon attenuation in the patient resulting in too low count values in the projections. The photon attenuation is generally a function of photon energy, body composition, density and on the source location. In non-homogeneous regions, such as the thorax, this makes it a complex radiation transport problem to solve. Since an iterative algorithm aims to find an image solution for which related projections converge to the measured projections, it is important that the model of the camera in the projector step also includes modeling of the attenuation.

3.4. Scatter contribution
The underlying reason for an unwanted scatter contribution in the image is that NaI(Tl) scintillation cameras do not measure imparted energy very well. The energy resolution (i.e. the statistical uncertainty in the energy signal) is about 10% (FWHM) at 140 keV. This means that a relative large energy window will be needed to maintain good counting statistics. If compton scattering occurs in the patient with only a slight change in energy and direction, there will be a chance that events from those photons are accepted within the energy window due to the energy resolution. These events are, however, to some degree miss-placed resulting in a decreased images contrast. Quantitative imaging is also difficult to perform since the scatter contribution and its spatial distribution is non-linear.

3.5. Collimator response and septal penetration
The underlying design of a parallel collimator is the main reason for the poor spatial resolution in scintillation camera imaging. Since photons initially are emitted in an isotropic direction, then the principle for geometrical collimation imply that a narrow collimator hole (better spatial resolution) will lead to fewer detected counts (a lower sensitivity) for a given activity and acquisition time. The poor spatial resolution also makes it difficult to quantify activity concentrations in small volumes. This limits the accuracy, for example, regarding tumor dosimetry. Furthermore, collimators are optimized for particular photon energies. This introduces a problem if additional photons with higher energies than the principal energy are emitted during the decay. For these photons, the collimator septa might not be sufficiently thick to guarantee a high fraction of photon absorptions. The results will be septum-penetration with a result in a point-spread function of long wide tails and characteristic streaks that match the flat sides of the hex-holes. The result in the image will be a lower contrast and counts outside the patient.

4. Corrections needed for therapy imaging
Today, most quantitative correction methods are implemented in some kind of iterative reconstruction method since the projector/backprojector is a natural stage in implementing the degrading effects that occur in real measurements. It should be remembered that the reconstruction processes try to find a distribution for which its calculated projections are as close as possible to the real measurements. Implementation of physical effects in the projector/backprojector will, in reality, mean a correction for the effects. Monte Carlo simulations are very important when validating correction methods because it is possible to simulate images of a clinically realistic activity distribution without any photon interactions in the phantom. Thus, those images will represent the case of perfect correction methods and therefore are very useful as reference images. Measurements using physical phantoms are also possible for comparison but the problems with this type of measurements are in the somewhat limited flexibility in making clinical realistic images. Furthermore, realistic physiological movements with a high degree of realism can be hard to implement on a physical phantom.

4.1. Compensation for photon attenuation
Compensation for attenuation is relatively easy to implement in the forward projector. Today, usually such information is obtained from a registered CT study. One problem here is to properly convert the data from the Hounsfield voxel numbers (which is obtained from a spectrum of x-ray photons) to linear attenuation coefficients that reflect discrete photon energies. Another potential problem today,

with modern hybrid SPECT/CT systems is that images obtained from a fast CT acquisition do no always reflect the 'average' attenuation caused by the patient's breathing. It should be remembered that SPECT data is acquired over a long time period in which the patient breathe. Therefore, an attenuation map should reflect the average of the respiratory motion and not a snapshot.

4.2. Compensation for scatter

Scatter correction is somewhat more complicated to compensate for as compared to attenuation. Monte Carlo simulations have been shown to be very useful to understand the scatter characteristics (qualitatively and quantitatively) since it is impossible to properly measure only scatter in the main energy window. Most compensation methods are based on either acquisition of scatter estimates from multiple energy windows or modeling scatter distribution, analytically, directly in the reconstruction algorithm. The most common energy-window method is probably the triple-energy window method where two narrow energy windows are placed on each side of the main energy window and a scatter images is obtained from a weighted average of the two related scatter images. Energy-window based methods are relative easy to implement by do not always reflect the scatter in the main peak properly because events from photon scattered with different number of orders appear in different parts of the measured energy spectrum. Most of the scatter in the main energy window usually comes for single-order scattered photons that have been scattered in very narrow angles and therefore only loose a small fraction of their energy. To reduce noise problems, scatter window data can be implemented in an iterative algorithm as an additive component rather than be subtracted from the SPECT projections prior to reconstruction. The ESSE method [5] is an example of an algorithm where Monte Carlo simulated scatter kernels as function of depth are used to model the overall scatter distribution. It is also possible to implement full Monte Carlo simulations to estimate the scatter. Such work has been published for ^{99m}Tc and ^{131}I [2]. These sophisticated approaches are expected to provide the best quantitative accuracy and include Monte Carlo simulation of compton scattering.

4.3. Compensation for collimator-response

The poor spatial resolution caused by the collimator can partly be compensated for by introducing a distance-depending blurring in the projector step. If no penetration occurs a gaussian-shaped distant-dependent model of the blurring can be used. However, septal penetration introduces non-symmetrical Point-Spread Functions (PSF) with distinct penetration streaks. A way to account for this is to use pre-Monte Carlo calculated PSFs using Monte Carlo simulations and store these kernels as a function of distance [4]. Monte Carlo simulations do also allow for calculations of the fraction of events in the image originated from photons penetrating the septa. This information can be very useful, particular in collimator design for radionuclides used for therapy. For ^{99m}Tc, these procedures with forced-detection of photons for fast collimator simulations has shown to be clinically practical but for more complex decay schemes, such as ^{131}I, it is also necessary to include scatter in the crystal and in the camera housing, and therefore these forcing methods may be difficult to be used.

5. Monte Carlo simulations of therapy imaging

Below, the examples for ^{111}In, ^{177}Lu, ^{131}I exemplify the unique possibilities with Monte Carlo simulations. A standard NaI(Tl) scintillation camera with both a MEGP and a HEGP collimator and with an energy resolution set to 10% (FWHM) at 140 keV was simulated. The XCAT anthropomorphic voxel-based computer phantom [13] was used. Figures 1-3 show in column (a) images from geometrically collimated photons that they have not interacted in the phantom or in the camera head (except for the crystal). Column (b) show images from attenuated photons but not scattered in the phantom. No events from septal-penetration or interactions in the camera are included. Column (c) show images from attenuated photons that may have penetrated the collimator septa and scattered in the crystal. Column (d) includes, in addition to the previous case, also photons scattered in the phantom. Finally, column (e) show images with all events including backscatter. Poisson noise has been added to all images in order to reflect a more clinical realistic image.

5.1. ^{177}Luthetium imaging

This radionuclide, labelled to peptides such as Lu-(DOTA0,Tyr3) octreotate, have demonstrated successful treatment of neuro-endocrinal tumours [6]. The decay includes some photon energies suitable for imaging. The two most important photon energies are 208 keV (11%) and 113 keV (8%). The ^{177}Lu decay is relatively clean with only one photon emitted with an energy that is higher than the principal energy. The effect of attenuation can clearly be seen in figure 1(b) compared to 1(a), the penetration and scatter do not affect the image quality so much, because the MEGP collimator is optimized for a larger energy than 208 keV. Thus, this radionuclide makes possible to do quantitatively accurate therapy imaging.

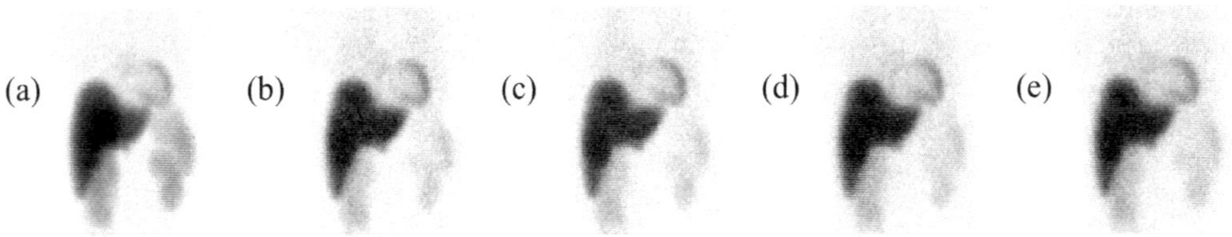

Figure 1. The images from ^{177}Lu photons, which are relatively sharp since few events come from septal-penetration and backscatter in the camera housing.

5.2. ^{131}Iodine imaging

^{131}Iodine has been used for many years in therapy of the thyroid and recently also labeled to antibodies, such as the ^{131}I-tositumomab for treatment of lymphomas. The ^{131}I decay includes a photon with the energy of 364 keV (80%) which makes therapy imaging possible. However, there is also emission of additional photons of higher energies which significantly affect the image quality. The two most important energy are 637 keV (7.3%) and 722 keV (1.8%).

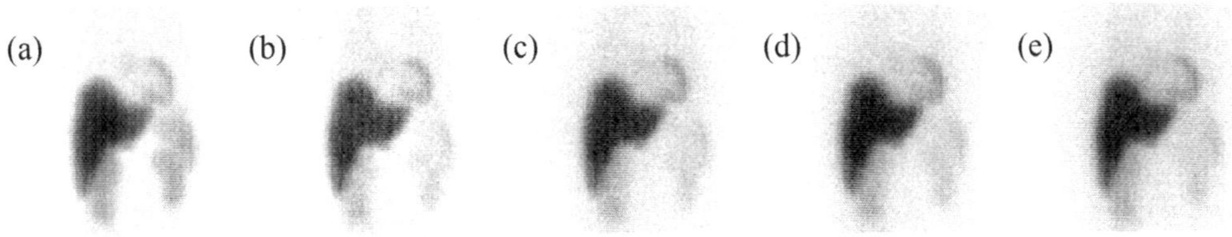

Figure 2. Images of ^{131}I showing some patterns caused by the collimator holes. The events from septal-penetrated (c) scattered photons and (d) backscatter photons (e) appear as a blurred effect on the image quality.

Dewaraja *et al.* have used the Monte Carlo method to characterize the images obtained from ^{131}I-labelled monoclonal antibodies [3]. Due to the complexity of the decay and the interaction within the camera, they have also investigated methods for a full Monte Carlo simulation of the scatter contribution implemented in an iterative reconstruction procedure [2]. The program SIMIND was used and the approach was that scatter, estimated in form of simulated images, is calculated at different iterations by exporting the estimate of the tomographic image, together with the patient's registered CT image to the Monte Carlo program, to perform an extended simulation including penetration in the collimator and backscatter in a compartment behind the crystal.

5.3. ^{90}Yttrium imaging

This radionuclide is a candidate for radionuclide treatment due to its large β-particle energy (and related particle range). One problem is that no γ-photons are emitted in the decay. However, during the β-interaction with surrounding tissues, bremsstrahlung photons can sometimes be emitted which can be detected by a scintillation camera. A major problem is, however, that bremsstrahlung photons spans in energy from the maximum β-particle energy down to very small energies. There is, thus, not a

Journal of Physics: Conference Series **317** (2011) 012016 doi:10.1088/1742-6596/317/1/012016

distinct photo-peak in the energy spectrum. This implies difficulties when defining which events should be regarded as 'good' events. Regardless of energy window position, there will always be a mixture of events coming from photon scattered in the patient, camera crystal and housing, collimator, PMTs etc. together with events from photons that have penetrated the collimator septa. Scatter correction using multiple-energy windows are, therefore, not appropriate. The probability for bremsstrahlung generation per decay is also small leading to noisy images. In our application of bremsstrahlung imaging at the Lund University, 'good' events are defined as those resulted from a photon that was not scattered anywhere and that passed the collimator hole geometrically. All other events are defined as 'bad' and then Monte Carlo calculated scatter- and collimator kernels are used for compensation within an iterative reconstruction procedure. A consequence of this definition was that the sensitivity could only be obtained from Monte Carlo simulations. Comparison with experimental measurements, however, concluded that a good activity recovery could be obtained [11].

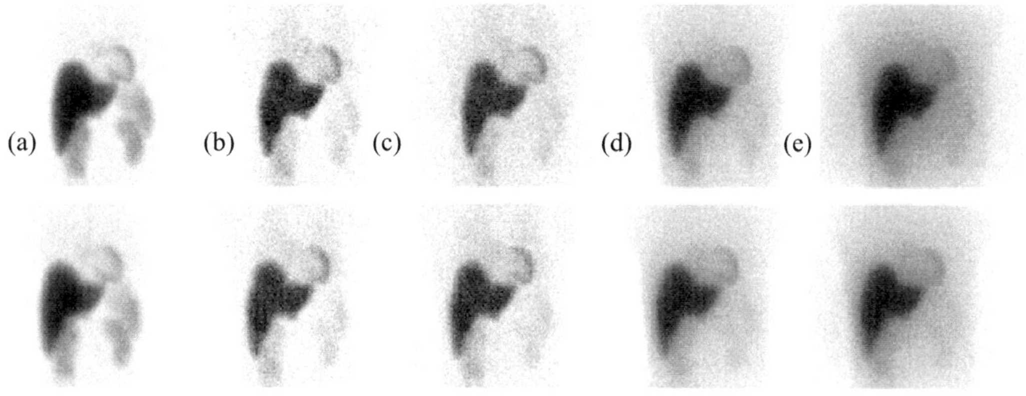

Figure 3. Simulated images of ^{90}Y bremsstrahlung photons. Upper row shows images simulated with the MEGP collimator and lower row shows images with the HEGP collimator. Note the penetration in (c) and the contribution from scatter in (d).

An implementation of bremsstrahlung imaging applied on data from an ongoing high-dose clinical trial with treatment of B-cells Lymphoma using ^{90}Y-labeled ibritumomab tiuxetan antibodies imaging has been described by Minarik *et al.* [10]. The administered activity needed to get the predicted absorbed dose to the main organ-at-risk was calculated from a pre-therapy study with ^{111}In-labeled ibritumomab tiuxetan antibodies. During treatments with ^{90}Y-labeled ibritumomab tiuxetan antibodies, several multiple SPECT/CT studies were made by bremsstrahlung imaging. The images were then quantified by correcting for non-homogeneous attenuation, scatter and collimator-response including septal penetration.

5.4. Characterization of events in the energy window

Monte Carlo calculations can help in characterizing the relative contribution to the images. The results in table 1 are from an example of such characterisation, taken from the simulations of such images. We can note that for ^{177}Lu most events comes from geometrically collimated photons and a small fraction from penetrations. This is because the collimator used is a medium-energy collimator with good absorption probability for 208 keV photons. No back-scatter occurs. However, a significant fraction of events comes from photons scattered in the phantom. For ^{90}Y, a large fraction comes from photons back-scattered behind the crystal especially for the MEGP collimator. This comes mainly from the high-energy bremsstrahlung photons that are highly penetrating through the collimator and crystal having enough energy to contribute to events in the energy window even though they have been backscattered with a large angle.

Table 1. Characterization of events, registered within the energy window, for four radiopharmaceuticals.

Type of event	^{177}Y	^{131}I	^{90}Y(ME)	^{90}Y(HE)
Geometrically collimated	68%	41%	13%	19%
Septal penetration	4%	19%	19%	10%
Scattering in the phantom	25%	17%	40%	49%
Backscattering behind the crystal	0%	9%	49%	32%

There is also a large fraction of events that comes from photons scattered in the phantom. However, this fraction is lower for ^{131}I than for ^{177}Lu because scatter is more easily discriminated for higher photon energies (increased separation between the full-energy peak and the Compton edge). For ^{131}I, there is also a fraction of events from photons penetrating the collimator. These events are most likely coming from the 637 keV and 722 keV photons that after penetration are compton scattered in the crystal before escaping and contributing to the main energy window with their partial energy deposition.

5.5. Characterization of events in the energy spectrum

It is not trivial to decide where energy windows for scatter corrections should be located when imaging a radionuclide with multiple photon emissions. Photons can be properly collimated but scattered with an escape in the crystal, leading to a 'good' registered event with lower energy. Since this event is correctly positioned, a subtraction is, therefore, not a good choice. By separating the events in categories, it is possible from a Monte Carlo simulation to calculate the relative contribution of 'good' and 'bad' events. Figure 4 and 5 show energy spectrums for each of the above discussed radionuclides. For display purposes, the information in figure 4 is limited to show the total spectrum corresponding to real measurements and the events from geometrically collimated photons that have not been scattered anywhere and, hence, represent 'good' events. For ^{177}Lu, it can be seen from figure 4(a) that scatter in the phantom is the major contributor to unwanted events in the main energy window. In the ^{131}I case, figure 4(b), it is evident that the high-energy photons, although theoretically with a low abundance, cause a lot of scattering and septal-penetration.

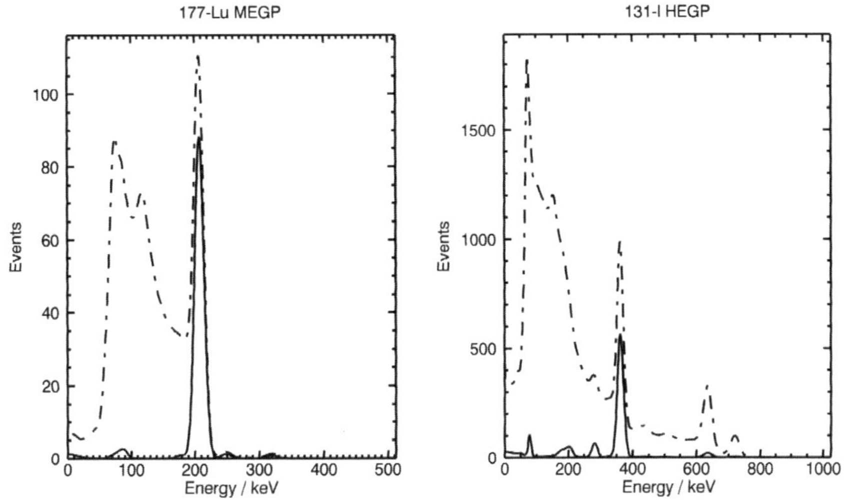

Figure 4. Examples of the ^{177}Lu (MEGP collimator) and the ^{131}I (HEGP collimator) energy spectrums, where the 'good' events are coming from unscattered photons; the geometrically collimated photons are plotted (solid) together with the total spectrum (dashed-dotted) representing a real measurement.

Figure 5(c) and 5(d) shows the ^{90}Y energy spectrum for the two collimators. There is much more penetration and scatter when using the MEGP than the HEGP collimator, but there is a relatively small increase in the geometrically collimated photons.

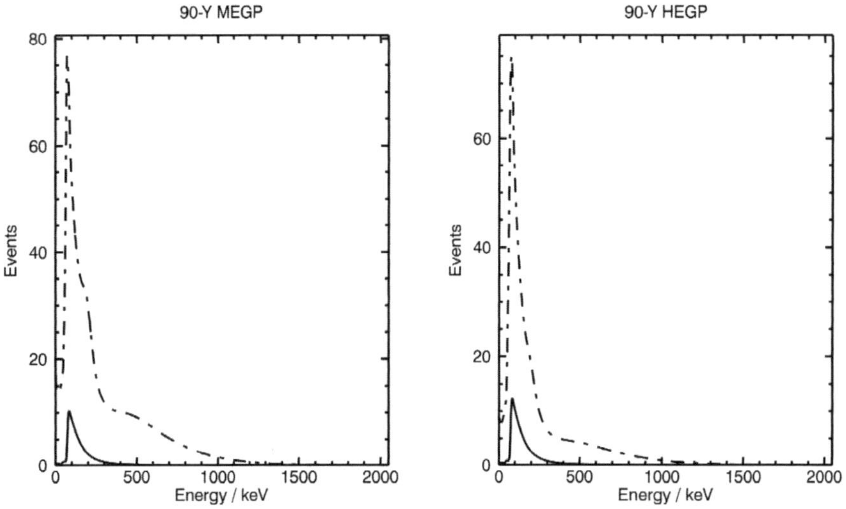

Figure 4. Examples of the ^{90}Y energy spectrums with the MEGP and HEGP collimators, where the 'good' events are coming from unscattered photons; the geometrically collimated photons are plotted (solid) together with the total spectrum (dashed-dotted) representing a real measurement.

7. Conclusion

The Monte Carlo method has shown to be very useful when evaluating imaging properties in both diagnostic and therapeutic nuclear medicine applications. In radionuclide therapy, most of the radionuclides used are not optimal for imaging and they may result in a large contamination of unwanted events, limiting the image quality and reducing the accuracy in quantitative measurements. However, by studying these radionuclides in an imaging situation using Monte Carlo methods, a deeper understanding of the degradation can be gained with a good possibility to develop effective correction methods for most of these effects.

References

[1] Briesmeister J F 2000 *MCNP - A General Monte Carlo N-Particle Transport Code* v. 4C (Los Alamos, NM: Los Alamos National Laboratory)
[2] Dewaraja Y K, Ljungberg M and Fessler J A 2006 *IEEE Trans. Nucl. Sci.* **53** 181-8
[3] Dewaraja Y K, Ljungberg M and Koral K F 2000 *J. Nucl. Med.* **41** 123-30
[4] Du Y *et al* 2002 *IEEE Trans. Nucl. Sci.* **49** 668-74
[5] Frey E C and Tsui B M W 1996 *Conf. Rec. IEEE Med. Imaging Conf.* (Anaheim, CA) pp 1082-86
[6] Garkavij M *et al* 2010 *Cancer* **116** 1084-92
[7] Jan S *et al Phys. Med. Biol.* **49** 4543-61
[8] Ljungberg M and Strand S E 1989 *Comp. Meth..Progr. Biomed.* **29** 257-72
[9] Ljungberg M, Strand S E and King M A 1998 *Monte Carlo Calculation in Nuclear Medicine: Applications in Diagnostic Imaging* (Bristol and Philadelphia: IOP Publishing)
[10] Minarik D, Sjogreen-Gleisner K, Linden O, Wingardh K, Tennvall J, Strand S E and Ljungberg M 2010 *J. Nucl. Med.* **51** 1974-8
[11] Minarik D, Sjogreen Gleisner K and Ljungberg M 2008 *Phys. Med. Biol.* **53** 5689-703
[12] Nelson R F, Hirayama H and Rogers D W O 1985 *The EGS4 Code System* (Stanford: SLAC)
[13] Segars W P, Sturgeon G, Mendonca S, Grimes J and Tsui B M 2010 *Med. Phys.* **37** 4902-15

Initial tests of a new phantom for investigation of spatial resolution, partial volume effect and detectability in nuclear medicine tomography

M Söderberg[1], U Engeland[2], S Mattsson[1], G Ebel[2] and S Leide-Svegborn[1]

[1] Department of Clinical Sciences, Medical Radiation Physics Malmö, Lund University, Skåne University Hospital, SE-205 02 Malmö, Sweden
[2] Scivis wissenschaftliche Bildverarbeitung GmbH, DE-37085 Göttingen, Germany

E-mail: marcus.soderberg@med.lu.se

Abstract. A new phantom has been designed that can provide simultaneously different target to background activity ratios with a linearly changing diameter of lesions. The purpose of the study was to describe and perform initial measurements with the phantom aimed to characterize different nuclear medicine tomographic systems and reconstruction algorithms in their performance and behaviour concerning partial volume effect (PVE) and detectability by varying the acquisition parameters and the count statistics. The phantom has an external vessel whose outline is half-cylindrical and allows it to be incorporated into an anthropomorphic thorax phantom. The phantom itself contains 16 fillable cones with an inner diameter linearly decreasing from 16 mm to 2 mm and a wall thickness of 1 mm acrylic glass. They as well as the outer vessel were separately filled with ^{99m}Tc- and ^{18}F-solutions respectively of different activity concentrations. The phantom was easy to fill and air bubbles could easily be avoided. Images taken using a SPECT/CT and a PET/CT system are presented as well as evaluations of PVE. The new phantom seems to be useful for comparison and optimisation of different acquisition and reconstruction parameters in nuclear medicine tomographic studies and for comparisons between various tomographic units.

1. Introduction

Nowadays, the majority of phantoms used to characterize a single photon emission computed tomography (SPECT) or positron emission tomography (PET) system are cylinders containing fillable inserts e.g. the Jaszczak PET/SPECT phantom [1] and NEMA IEC body phantom set™ [2], or phantoms consisting of multiple discrete disks or sheets e.g. the Hoffman phantom [3] and the porous phantom [4]. Tissue-equivalent anthropomorphic phantoms, as the RSD Alderson heart/thorax phantom, are also available [5]. A project named MADEIRA (Minimizing Activity and Dose with Enhanced Image quality by Radiopharmaceutical Administrations) cofunded by the European Commission within the Seventh Euratom Framework Programme aims to optimise the spatial resolution, the signal-to-noise ratio and reduce the radiation dose in 3D nuclear medicine imaging technologies [6]. One way to succeed is to find optimal conditions for different tomographic reconstruction algorithms in terms of number of counts, number of projections and matrix size.

Essentially for detection of lesions in SPECT and PET images is the contrast to noise ratio (CNR) of lesion to background. As there is always an intrinsic resolution volume defined by the camera, collimator, radionuclide, acquisition protocol and reconstruction method, the size of the lesion is also important. Below a certain volume, the reconstructed intensity tends to diffuse into neighbouring voxels, thereby resulting in very low target to background ratios (underestimation of activity concentration in the target) [7]. This effect is called partial volume effect (PVE) or as proposed by Skretting [8], intensity diffusion. To investigate the importance of this effect and to find methods to correct for it [9], a new patent filed phantom (European patent application no. 09008184, "Phantom for a tomographic medical imaging apparatus"), the "MADEIRA" phantom, has been designed so that it can simultaneously provide different target to background activity ratios with a linearly changing diameter of active or inactive lesions.

The purpose of the present study was to describe and perform initial measurements with this new phantom aimed to characterize different nuclear medicine tomographic systems and reconstruction algorithms in their performance and behaviour concerning PVE and detectability by varying the acquisition parameters and the count statistics.

2. Material and methods

2.1. The MADEIRA phantom

The phantom has an external vessel whose outline is half-cylindrical and allows it to be incorporated into an anthropomorphic thorax phantom [5]. The phantom itself contains 16 cones with an inner diameter linearly decreasing over a length of 19 cm from 16 mm to 2 mm (figure 1). The wall thickness of the cones is 1 mm. The 16 cones are separately fillable with activity in water solution as is the outer vessel. The phantom is constructed of acrylic glass for visual inspection of bubble free filling.

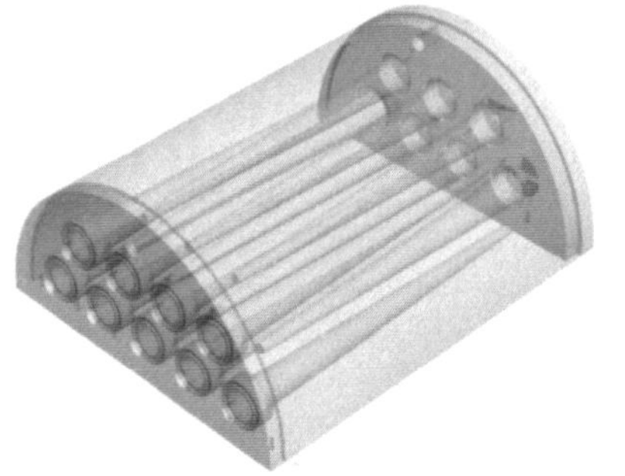

Figure 1. The construction of the MADEIRA phantom.

2.2. Phantom measurements

The concept was to minimize the number of necessary acquisitions to get a fairly good sampling of all acquisition parameters. At the beginning, we performed one SPECT and one PET measurement, respectively, with very good statistics. Our figures of merit are spatial resolution, PVE and detectability. Spatial resolution was evaluated as the full width at half maximum (FWHM) by drawing profiles across the center of the cones. PVE was evaluated by drawing profiles in the center along the length of the cones. Detectability was studied as a function of CNR and size using region of interests in and outside the cones.

2.2.1. SPECT measurements

The 16 cones were filled with ^{99m}Tc-solutions of different activity concentrations differing by a factor of ¾ from cone to cone; 8 cones above and 8 cones below background activity concentration. The lowest relative activity concentration was 0.1 and the highest 10.0, where 1.0 refers to the background activity concentration in the external vessel. The initial activity at the start of acquisition was 530 MBq. Acquisitions were performed with a Symbia T2 SPECT/CT (Siemens Medical Solutions,

Forchheim, Germany) in non-circular orbit step and shoot mode. The collimator used was a low energy ultra high resolution (LEUHR) [10]. The applied energy windows were a 15% photo-peak window symmetrically placed around 140 keV and a 15% lower scatter window. Each projection was acquired for 60 s and 240 projections were acquired over 360°. The acquisition matrix was 256×256. Images were reconstructed using the iterative algorithm ReSPECT 3.0; 8 iterations, nonlinear noise suppression and no Gaussian filter (Scivis wissenschaftliche Bildverarbeitung GmbH, Göttingen, Germany).

The number of projections was varied by taking only a subset out of the total projections, preserving an equidistant sampling of the angle space (120°, 80°, 60°, 48°, 40°, 30°, and 24°). The total number of photons was varied by Monte Carlo simulations (increasing Poisson noise), intended to virtually simulate shorter acquisition time or activity. The matrix size was varied by down sampling the projection matrix (256, 128, 64). This will make it possible to investigate the influence of count statistics, number of angles and matrix size simultaneously by just one (real) acquisition. All manipulations were carried out by using the Scivis scientific visualizer software, producing correct DICOM datasets that can be reconstructed by any DICOM capable software.

2.2.2. PET measurements

The 16 cones were filled with ^{18}F-solutions of different activity concentrations differing by a factor of ¾ from cone to cone. The lowest relative activity concentration was 0.1 and the highest 10.0, where 1.0 refers to the background activity concentration in the external vessel. The initial activity at the start of acquisition was 145 MBq. Acquisitions were performed with a Gemini 16 PET/CT (Philips Healthcare, Best, The Netherlands) using matrix size of 144×144 and a scan time of 15 minutes per bed position. The iterative reconstruction method used was LOR-TF-RAMLA ("BLOB-OS-TF") with default settings [11-12].

3. Results

The phantom was easy to fill using a syringe with a long needle and air bubbles could easily be avoided. The following two subsections will present a selection of the achieved results.

3.1. SPECT measurements

Figure 2 shows representative SPECT images of the MADEIRA phantom. Evaluation of PVE in the five cones with the highest activity concentration relative to the background activity concentration is shown in figure 3.

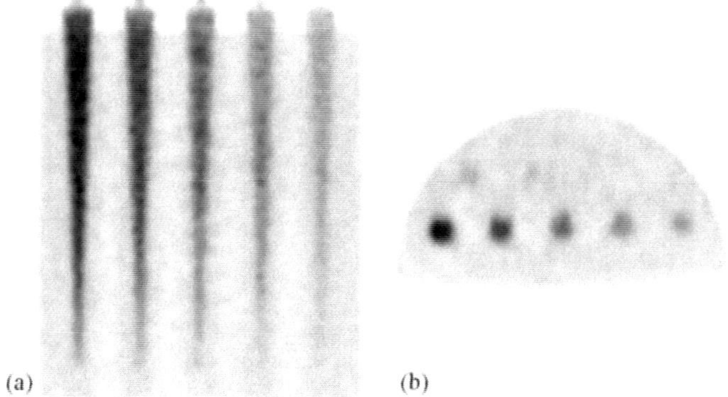

Figure 2. (a) Coronal and (b) axial SPECT images of the MADEIRA phantom (120 projection angles, matrix size 256, full statistics).

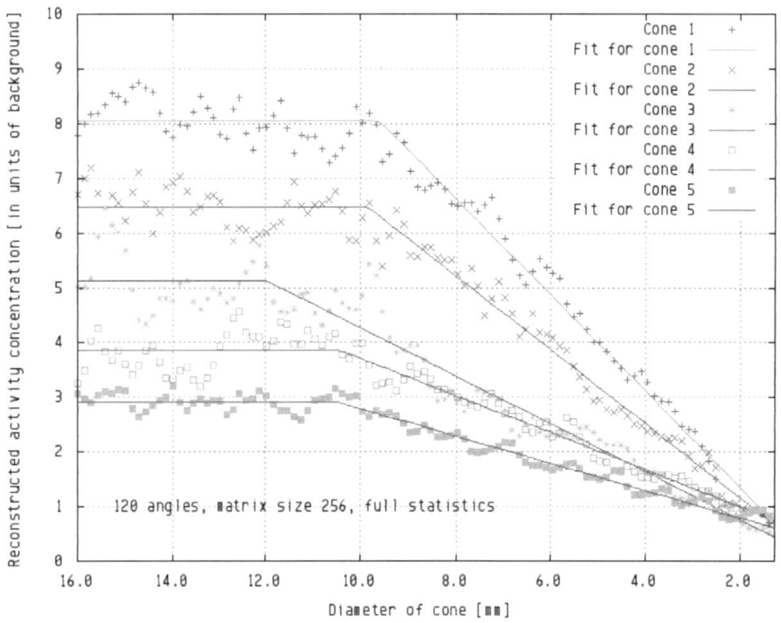

Figure 3. Reconstructed activity concentration normalized to the background activity concentration (measured in a cone-free area inside the vessel) as a function of the diameter of the cone.

Figure 4 shows the reconstructed activity concentration in the constant part as a function of the number of projections used for reconstruction. Figure 5 shows the diameter of the cone, where PVE starts to vary, with the number of projections used for reconstruction.

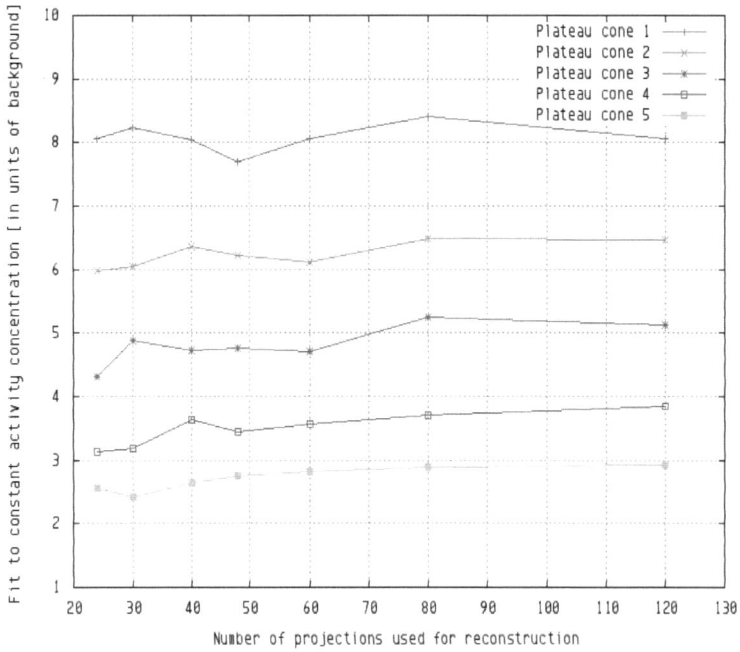

Figure 4. Reconstructed activity concentration in the constant part as a function of the number of projections used for reconstruction.

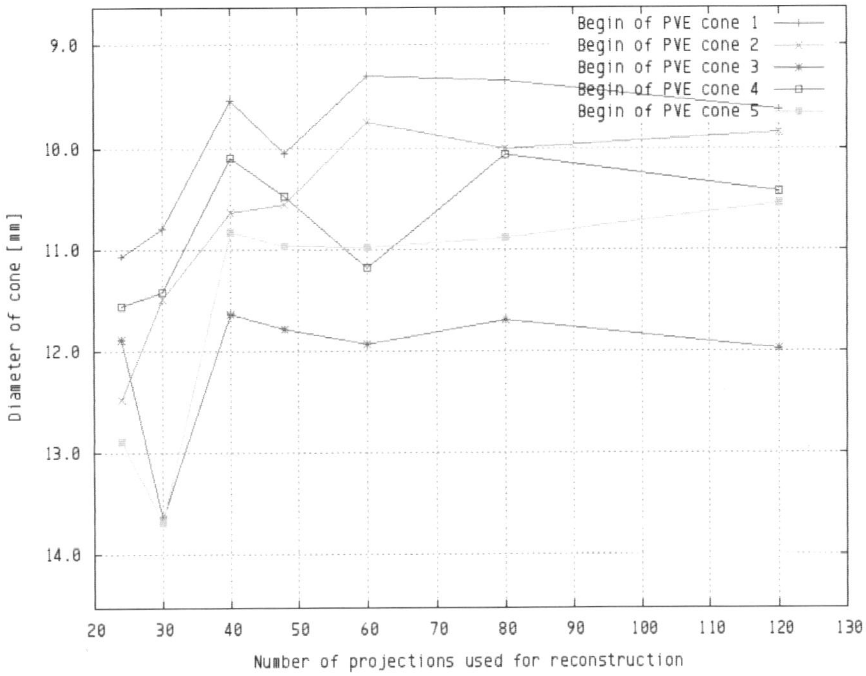

Figure 5. Diameter of the cone, where PVE starts, as a function of the number of projections used for reconstruction.

3.2. PET measurements

Figure 6 shows representative PET images of the MADEIRA phantom. Evaluation of PVE in the cone with the highest activity concentration relative to the background activity concentration is shown in figure 7.

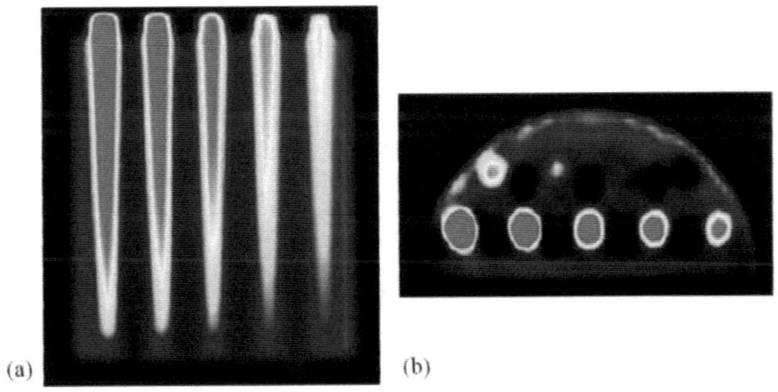

Figure 6. (a) Coronal and (b) axial PET images of the MADEIRA phantom.

4. Discussion

There are different available phantoms to characterize a SPECT or PET system [1-5]. The phantoms have various advantages and disadvantages, e.g. cost, levels of contrast and modelling of anatomy. The MADEIRA phantom, used in this study, has been designed so that it can simultaneously provide different target to background activity ratios with a linearly changing diameter of active or inactive lesions. To make measurements with the MADEIRA phantom under realistic clinical conditions, the outer vessel was designed to fit into the RSD Alderson heart/thorax phantom as an alternative to its lung inserts. A special baseplate to the Alderson phantom is under construction and additional measurements will be performed with the MADEIRA phantom in the thorax phantom.

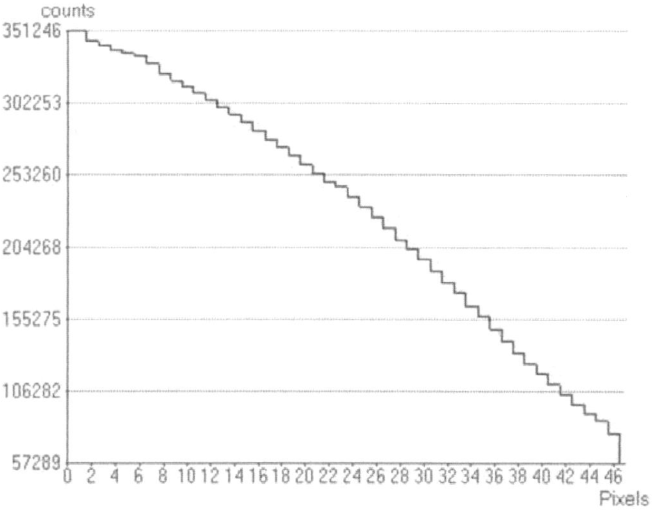

Figure 7. The number of counts as a function of the length (pixels) of the cone with the highest relative activity concentration.

Using the MADEIRA phantom, it is possible to characterize different nuclear medicine tomographic systems and reconstruction algorithms in their performance and behaviour concerning PVE and detectability. Evaluation of PVE for our SPECT measurements was presented in figure 3. As shown in figure 4, the level of reconstructed activity concentration in the constant part was not affected by the number of projections used for reconstruction, but by the diameter of the cone where PVE starts (see figure 5). Consequently, we found that a reduction of the number of projections will increase the importance of PVE.

Figure 6b showed possibilities using the MADEIRA phantom to demonstrate detectability. It is shown that the size of the uptake is reduced with lower activity concentration. Our evaluation of PVE for the PET images did not found any plateau before the starting point of the PVE (see figure 7). This finding was not expected and indicates that the cones should be constructed with larger base diameters. It also shows the shortcomings with the PET system. Normally, we would have expected a better spatial resolution for the PET system compared to the SPECT system. This finding shows the importance of the reconstruction process and it may be due to the chosen noise suppression and regularization mechanisms included in the reconstruction algorithms.

The cone walls were fabricated to be as thin as possible (1 mm) to minimize the "wall effect". Ideally, the cone walls should have non-zero thickness to prevent the fact that a shell of zero activity separates the cone solution and the background solution. This may affect quantitative measurements and evaluation of lesion detection, but it is believed to be neglected concerning evaluation of PVE.

The phantom used in this study is a prototype and improvements (positioning of cones, larger diameter of cone base etc.) will be performed before the phantom will be commercially available in the middle of 2011. It is not possible to disassemble the phantom and this can bring problems with mould due to lack of ventilation. We used pressure air to completely dry the phantom after use. The next generation of the phantom will improve the ease of cleaning (e.g. bigger emptying valves).

5. Conclusion

After minor improvements, the MADEIRA phantom has the potential to be a useful and an important practical tool for comparison and optimisation of different acquisition and reconstruction parameters in nuclear medicine tomographic studies and to find the best working point of a given system as well as for comparisons between various tomographic units.

Journal of Physics: Conference Series **317** (2011) 012017 doi:10.1088/1742-6596/317/1/012017

Acknowledgements
The work was carried out within the Collaborative Project "MADEIRA" (www.madeira-project.eu), cofunded by the European Commission through EURATOM Seventh Framework Programme (Grant Agreement FP7-212100).

References
[1] Jaszczak PET/SPECT Phantom (www.coneinstruments.com)
[2] Data Spectrum Corporation NEMA IEC Body Phantom Set™ (www.spect.com)
[3] Hoffman E J, Cutler P D, Digby W M and Mazziotta J C 1990 *IEE Trans. Nucl. Sci.* **37** 616-20
[4] DiFilippo F P, Price J P, Kelsch D N, Raymond J and Muzic F 2004 *Med. Phys.* **31** 1183-94
[5] RSD Alderson heart/thorax phantom (www.rsdphantoms.com)
[6] Hoeschen C, Mattsson S, Cantone M C, Mikuz M, Lacasta C, Ebel G, Clinthorne N and Giussani A 2010 *Rad. Prot. Dosim.* **139** 250-3
[7] Hoffman E J, Huang S C and Phelps M E 1979 *J. Comput. Assist. Tomogr.* **3** 299-308
[8] Skretting A 2010 *Eur. J. Nucl. Med. Mol. Imaging* **36** 536-7
[9] Rousset O G and Zaidi H 2006 *Quantitative Analysis in Nuclear Medicine Imaging* (Singapore: Springer) 236-71
[10] Siemens Medical Solutions 2005 *Symbia TruePoint SPECT/CT System Specifications* (www.medical.siemens.com)
[11] Hu Z, Wang W, Gualtieri E E, Hsieh Y L, Karp J S, Matej S, Parma M J, Tung C H, Walsh E S, Werner M and Gagnon D 2007 *IEEE Nucl. Sci. Symp. Conf. Rec.* **6** 4415-18
[12] Surti S, Kuhn A, Werner M E, Perkins A E, Kolthammer J and Karp J S 2007 *J. Nucl. Med.* **48** 471-80

Journal of Physics: Conference Series **317** (2011) 012018 doi:10.1088/1742-6596/317/1/012018

Monte Carlo simulations to assess differentiation between defects in cardiac SPECT

I Chrysanthou-Baustert[1], Y Parpottas[1,2], O Demetriadou[3], S Christofides[4,5], Ch Yiannakkaras[4], D Kaolis[4,5] and M Wasilewska-Radwanska[6]

[1] Frederick Research Centre, Nicosia, Cyprus
[2] General Department (Physics-Math), Frederick University, Nicosia, Cyprus
[3] Nuclear Medicine Department, Limassol General Hospital, Limassol, Cyprus
[4] Medical Physics Department, Nicosia General Hospital, Nicosia, Cyprus
[5] Biomedical Research Foundation, Nicosia, Cyprus
[6] AGH University of Science and Technology, Krakow, Poland

E-mail: isabelle.chrysanthou@gmail.com

Abstract. Differentiating between various types of lesions in nuclear cardiology is a challenge. This work assesses the level of differentiation achievable between various low contrast lesions, as encountered in nuclear cardiology. The parameters investigated are defect extend, defect thickness and perfusion reduction of the defect. The images have been obtained through Monte Carlo Simulations with the program SIMIND. Results show that acceptable size resolution is obtained for defects with an extend over 25x25mm. When thickness and perfusion reduction are both unknown, the imaging results are confounding. In this work, thickness and perfusion reduction cannot be differentiated. If one of the variables is known (thickness or perfusion reduction), imaging results can differentiate between the other unknown variable.

Introduction

The detection of a cardiac defect relies on resolving a low radiopharmaceutical concentration within a higher activity. Previous publications have assessed the limitations of resolving cardiac defect characteristics with simulations of a dynamic cardiac phantom [1-2] and with various commercially available post-processing softwares [3-4]. They report statistically relevant variations between the severity results computed by commercially available packages. The simulation studies further investigate the variation of minimum defect detection size with perfusion reduction and the inability to discriminate between different defect thicknesses and perfusion reductions in the heart. In this paper, Monte Carlo simulations are used to assess differentiation between low contrast defects, as encountered in nuclear cardiology. Three different parameters contributing to the resolution of lesions in the myocardium are analysed. The parameters investigated are defect extend, defect thickness and perfusion reduction of the defect (that is, the reduction of radiopharmaceutical concentration in the defect relative to the surrounding healthy heart).

Method

Monte Carlo simulations with the SIMIND [5] code were performed with the following settings: LEHR collimator, 140 keV photons with an energy resolution of 10%, a 20% energy window centred around 140 keV, a 64x64 matrix, a pixel size of 0.63 cm, 2 projections were recorded at 0° and 90°.

Journal of Physics: Conference Series **317** (2011) 012018 doi:10.1088/1742-6596/317/1/012018

Thirty-three simulations were run with various defects immersed in a radiopharmaceutical concentration. The defect sizes are shown in **Table 1**. The defects were sequentially placed in a container of length 16 cm, width 6 cm and thickness 2 cm filled with ^{99m}Tc (100%). The shape was chosen to simulate the cardiac wall. The defect extend (30x30, 25x25, 20x20, 15x15 and 10x10 mm^2) and thickness (5, 10, 15 mm) were varied. Further, the relative concentration of radiopharmaceutical (0%, 30%, 60%) was varied to simulate severe, moderate and mild perfusion reduction in the defects, respectively.

Table 1. The simulated defects sizes are indicated with a tick (√). For each of the indicated defect size, the severe, moderate and mild perfusion reduction was simulated.

Extend (mm)	Thickness (mm)		
	5	10	15
10x10		√	
15x15	√	√	√
20x20	√	√	√
25x25		√	
30x30	√	√	√

The analysis software was written in MATLAB (The MathWorks). The projections at 0° and 90° were analysed. Profiles along the x-axis, the y-axis and the z-axis were extracted from the projections. For quantitative analysis, the signal-drop, seen in the profiles due to the defects, was normalized to 100, and then inverted, which turned the signal-drop intensity into the signal intensity to increase. Then a gaussian function was fitted to characterize the amplitude and the FWHM of the signal intensity profile.

Results

All the defects could be observed by visual inspection. In figure 1, the 0° and 90° projections together with their profiles are shown for defect size of 25x25x10 mm and severe perfusion reduction (i.e. 0% radiopharmaceutical uptake in the defect), while in figure 2, are shown for defect size of 10x10x10 mm and severe perfusion reduction.

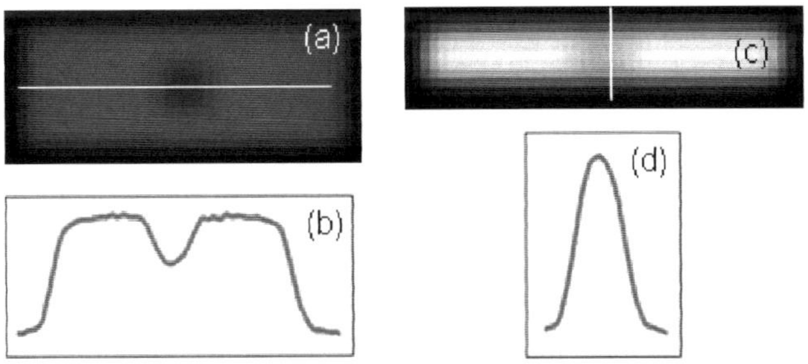

Figure 1. Simulation for defect size of 25x25x10 mm and severe perfusion reduction. (a) Image for 0° projection, and (b) the corresponding profile along the long axis (white line in a). (c) Image for 90° projection, and (d) the corresponding profile along the short axis (white line in c).

The quantitative analysis of the gaussian function, characterizing the inverted signal-drop intensity due to the defects, involves the computation of the FWHM and the amplitude. To visualise all the results, a scatter plot is shown in figure 3. The size is represented by the increasing square size and ranges from 1x1, 1.5x1.5, 2x2, 2.5x2.5 to 3x3 cm. The colour represents the perfusion reduction: green is mild, red is moderate and black is severe. The mark inside the square represents the thickness: ● is 5mm, + is 10mm and * is 15mm.

In a first step, the ability to resolve the extend of the defects is assessed. It can be observed from figure 3 (right part of the graph) that the defects with extend 30x30 mm are well resolved. The standard deviation of the fitted FWHM is of the order of 2.3%. The same can be argued for defects of extend 25x25 mm for which the standard deviation of the FWHM is of the order of 3.8%. For smaller defects, there is major overlap between all the defect extends. For defects with extend 20x20 mm^2 and below, the standard deviation of the FWHM is of the order of 10%.

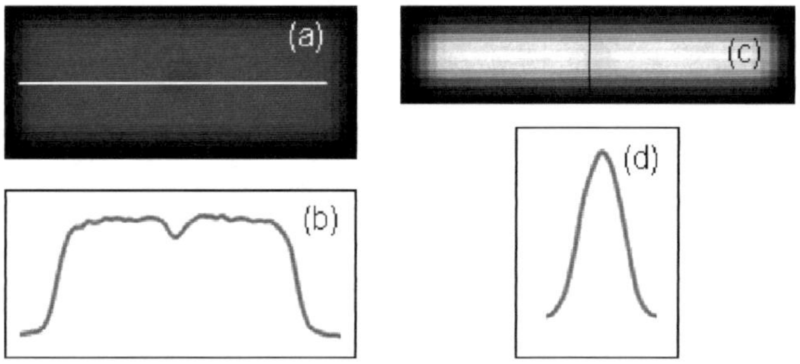

Figure 2. Simulation for defect size of 10x10x10 mm and 0% radiopharmaceutical uptake. (a) Image for 0° projection, and (b) the corresponding profile along the long axis (white line in a). (c) Image for 90° projection and (d) the corresponding profile along the short axis (black line in c).

In figure 3, it can be observed for the well resolved large-size defects (higher values of FWHM), that for each perfusion reduction (represented by the colour), there is differentiation between the thicknesses of the defects (represented by the internal mark), that is, thicker defects have larger amplitude in the figure. It can also be seen for the large-size defects that for each thickness, there is differentiation between the perfusion reductions of the defects, that is, more severed defects have larger amplitude. However, there is overlap between the thickness and the perfusion reduction of the defects. A thin defect with severe perfusion reduction produces the same signal-drop as a thick defect with moderate perfusion reduction. Only the two extreme cases (thick defect with severe perfusion reduction and thin defect with mild perfusion reduction) are well resolved.

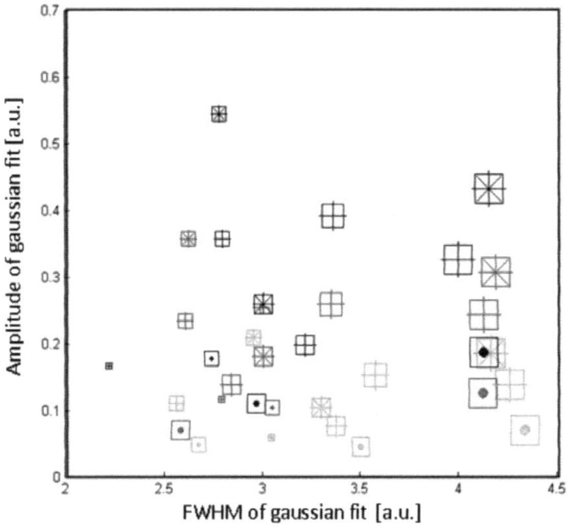

Figure 3. Scatter plot of the gaussian fit parameters, amplitude and FWHM, on the y and x- axis, respectively, for the three variables: extend of lesion, thickness of lesion and perfusion defect severity. The size is represented by the increasing square size and ranges from 1x1, 1.5x1.5, 2x2, 2.5x2.5 to 3x3 cm. The thickness is represented by the mark inside the square: • is 0.5 cm, + is 1 cm and * is 1.5 cm thickness. The perfusion defect severity is shown by the colour: black is severe, red is moderate and green is mild defect.

For the small defect sizes, even though it is still obvious the relation between (a) same size and thickness of defects with different severity and (b) same severity and sizes of defects with different thicknesses, as for the large-size defects, there is no correlation among defect sizes; a major overlap is observed.

The profiles along the short axis (figure 1d and 2d) do not exhibit signal-drop intensity due to the defects geometry with respect to the short axis. To analyse these data, a gaussian was fitted to these profiles for various thicknesses and perfusion reduction of the defects. The FWHM variations are smaller than 2.5%. All 90° projection profiles for the 30x30 mm are shown in figure 4. For a given extend of the defects (i.e. 30x30 mm), the profile amplitudes vary as follows: lowest amplitude for the thick defect with severe perfusion reduction, and highest amplitude for the thinnest defect with the mildest perfusion reduction. Overlap also exists for defects with various thicknesses and perfusion reduction.

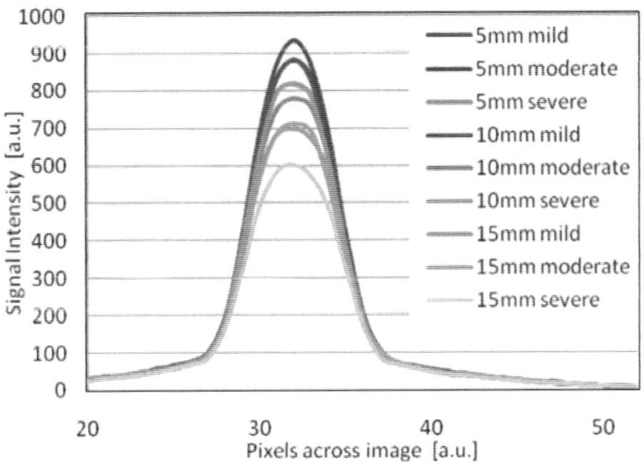

Figure 4. Profiles for 90° projections across the height of the simulated container with 30x30 mm inserted defects varying in perfusion reduction and thickness.

4. Conclusions

The difficulties in differentiating among different types of lesions in the heart with specialized post-processing software have been reported in previous publications. According to this investigation, the reason behind this problem is due to the basic limitations such as the low spatial resolution. Defect extend resolution becomes reliable above 25x25 mm. Even if the extend is well resolved for larger defect sizes, the thickness and perfusion reduction remains unresolved for any type of defect. It is therefore important to understand the clinical relevance of thickness and perfusion reduction in heart defects. In some clinical investigations, the physician might be interested to investigate only the perfusion reduction for a certain defect. Therefore, a constant defect thickness will be assumed and the report will focus on perfusion reduction of the defects.

The commercially available softwares are dealing with confounding imaging data (i.e. unresolved defect extends, defect thicknesses and perfusion reduction) and it is therefore not surprising that differences in various commercial implementations are reported.

Acknowledgments

This work is funded by the Cyprus Research Promotion Foundation and the European Regional Development Fund through the project ΥΓΕΙΑ/ΔΥΓΕΙΑ/0308/11: Optimising Diagnostic Value in SPECT Myocardial Perfusion Imaging.

References

[1] Bartosik J *et al* 2006 *Clin. Physiol. Funct. Imaging* **26**(6) 343-350
[2] El-Ali H *et al* 2005 *Nucl. Med. Commun.* **26**(12) 1125
[3] Wolak A *et al* 2008 *J. Nucl. Cardiol.* **15**(1) 27-34
[4] Guner L *et al* 2010 *Eur. J. Nucl. Med. Mol. Imaging* **37** (11) 2070-2078
[5] Ljungberg M 1997 The SIMIND Monte Carlo program *http://www.radfys.lu.se/simind/*

Journal of Physics: Conference Series **317** (2011) 012019 doi:10.1088/1742-6596/317/1/012019

Design of heart defects for in vitro perfusion reduction testing

I Chrysanthou-Baustert[1], Y Parpottas[1,2], O Demetriadou[3], D Kaolis[4,5], S Christofides[4,5], Ch Yiannakkaras[4] and M Wasilewska-Radwanska[6]

[1] Frederick Research Center, Nicosia, Cyprus
[2] General Department (Physics-Math), Frederick University, Nicosia, Cyprus
[3] Nuclear Medicine Department, Limassol General Hospital, Limassol, Cyprus
[4] Medical Physics Department, Nicosia General Hospital, Nicosia, Cyprus
[5] Biomedical Research Foundation, Nicosia, Cyprus
[6] AGH University of Science and Technology, Krakow, Poland

E-mail: isabelle.chrysanthou@gmail.com

Abstract. Solid heart defects were designed to simulate regions of various contrast agent concentration by simply being immersed in a container with a radiopharmaceutical solution. The design was tested and verified by Monte Carlo simulation. Preliminary experimental results were taken on a gamma camera. The lower radiopharmaceutical concentrations were obtained by removing a certain volume of material, isotropically, from a solid cube of perspex material. The resolution of the imaging modality does not resolve the structure of the cube and the carved cube appears as a region of lower radiopharmaceutical concentration. It is aso verified with Monte Carlo simulations, obtaining profiles across low radiopharmaceutical concentration areas that vary by less than 3%. The preliminary imaging results confirmed that the gamma camera could not resolve the internal structure of the carved cube and that partial volume effect resulted in image intensity reduction.

1. Introduction

Quantitative analysis of extent and perfusion reduction of heart problems in nuclear cardiology is calculated by processing the raw SPECT images with commercial software [1-3]. The verification of these results can only be made by comparing the results to a priori know situations usually with the use of a phantom. While extend is easy to be simulated in a phantom, perfusion reduction can only be tested by introducing a small container filled with a lower radiopharmaceutical concentration within a higher radiopharmaceutical concentration. We have designed solid defects, which when immersed in a solution of high radiopharmaceutical concentration, produce regions of known lower radio-pharmaceutical concentration. In this paper, the design of the solid heart defects, together with its validation using Monte Carlo simulations, and preliminary imaging results with a clinical scanner (γ-camera) are presented.

2. Method

The new defect design consists of a solid cube from which a certain volume has been removed in an isotropic way such that radiopharmaceuticals can penetrate the perforated structure but the resolution of the imaging modality can not resolve the structure of the cube. As a result of the partial volume effect, a predictable lower radiopharmaceutical concentration appears in the image.

2.1. Defects using Monte Carlo simulations
Monte Carlo simulations with the SIMIND code were performed to investigate the validity of the heart defect design. The following settings were used: LEHR collimator, 140 keV photons with an energy resolution of 10%, a 20% energy window centered around 140 keV, a 64x64 matrix, a pixel size of 0.63 cm, and one projection at 0° was recorded. The simulation setup is described below.

First, a rectangular source of 6x4x2 cm size was simulated with a cubic insert of 10 mm sides. The insert was programmed to contain various lower concentrations of radiopharmaceutical (32%, 50% and 66.5% of the surrounding source). This reduction simulated a defect with reduced perfusion.

Second, a similar setup was simulated with a solid cube with a certain percentage of its material removed was placed at the position of the cubic insert. The material was removed in such a way that the clinical scanner could not resolve its structure and, thus, the simulated partial volume effect, where the overall radiopharmaceutical concentration seen by the clinical scanner, was reduced by the amount of the solid volume remaining. Various radiopharmaceutical concentrations were simulated by increasing or reducing the volume of the removed material for the solid cube. In each cube, 16 small cylinders were removed with diameter 1.6, 2 and 2.3 mm resulting in a percentage of 32%, 50% and 66.5% removed volume, respectively.

2.2. Manufactured defects
Small volumes of perspex were cut and a predefined volume of material was removed in such a way that water can penetrate through the remaining structure and that the gamma camera cannot resolve its fine structure. As a result, the gamma camera detects a lower radiopharmaceutical concentration.

Perspex cubes of 10 mm side were cut on an 80 W CO_2 laser cutter. 52% of its volume was removed by cutting out 23 small cylinders of 1.7 mm diameter. Figure 1 shows a manufactured defect of 10x10x10 mm with 52% of material removed.

We used coloured water to assess the ability of the holes to fill up with water without any remaining air bubbles when immersed. The capillary effect helped to fill the holes without any air bubbles.

For the in vitro experiments, two similar cubes were carefully immersed into a water cylindrical test flask container, of 1.5 cm diameter and 8 cm length, to which radiopharmaceutical was added. The vial was tightly closed and inserted into the Nuclear Associates PET/SPECT Phantom Source Tank (Fluke Biomedical). SPECT imaging was performed on a GE advantage scanner.

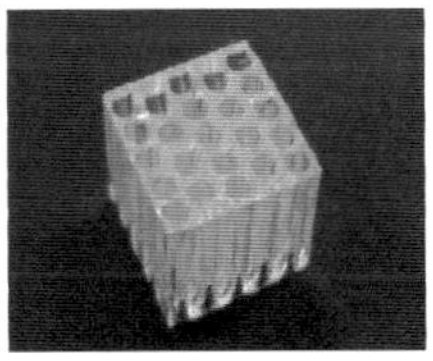

Figure 1. Manufactured perspex defect of 10x10x10 mm with 52% of material removed.

3. Resullts
3.1. Monte Carlo simulations
The results of the Monte Carlo simulations were assessed visually by drawing profiles across the regions of the two types of simulated defects (a vial with 50% radiopharmaceutical, a cube with 50% of its volume removed as small cylinders). As figures 2 and 3 show, there is no visual difference between the images and, further, there is perfect agreement between the two different defect types with deviations in profiles lower than 3%. The internal structure of the cubes could not be resolved by the gamma camera.

On the 0 ° projections profiles, a reduction in counts is observed in the region of the cubes with removed volumes. The image intensity dropped by approximately 13.5%. The internal structure of the

Journal of Physics: Conference Series **317** (2011) 012019 doi:10.1088/1742-6596/317/1/012019

cubes could not be resolved by the gamma camera. However, on the 90 ° projections profiles, no signal-drop could be observed in both cases because of the vial geometry.

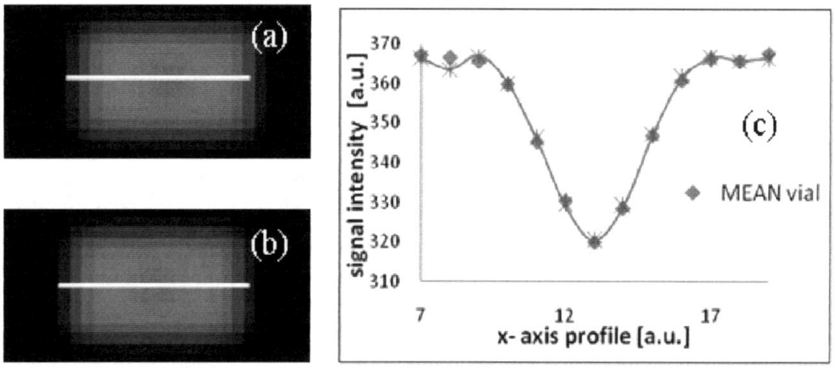

Figure 2. 0° projection of a source, with Monte Carlo simulations, containing (a) a vial with 50% radiopharmaceutical, (b) a cube with 50% of its volume removed as small cylinders, (c) profiles across the images (white lines); blue colour for the first case and red colour for the second case.

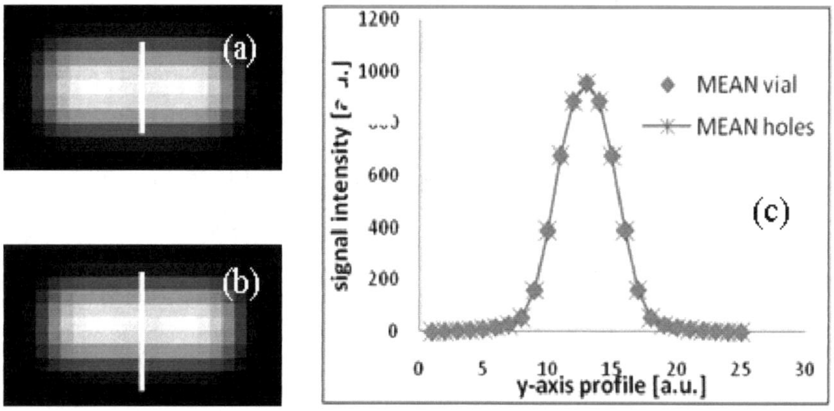

Figure 3. 90° projection of a source, with Monte Carlo simulations, containing (a) a vial with 50% radiopharmaceutical, (b) a cube with 50% of its volume removed as small cylinders, (c) profiles across the images (white lines); blue colour for the first case and red colour for the second case.

3.2. Manufactured defects

Figure 4 shows a screenshot of the results obtained from the structure (two similar cubes immersed in the container) described in section 2.2 using a gamma camera at 0° projection. In the region of the two cubes (top parts of figures 4a and 4b), fewer counts are observed. The count reduction in the region with the cubes is of the order of 14%.

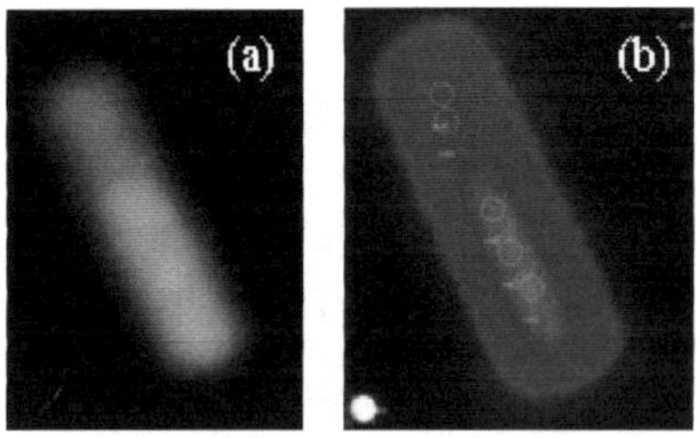

Figure 4. Screenshots of the results obtained from the structure described in section 2.2 using a gamma camera at 0° projection: (a) black and white scale, (b) colour scale with ROIs (white circles). The two circles (ROIs) at the top part of figure 4b include the two cubes.

Conclusions
The results show that various predefined radiopharmaceutical concentrations can be created in one single container without the need of further dilutions. The experimental setup can be easily reproduced. This concept could potentially be used to design new generation of QA phantoms in assessing the accuracy of the image reconstruction algorithms, in the quantification of radiopharmaceuticals and in the assessment of the limitation of quantitative nuclear medicine.

Acknowledgments
This work is funded by the Cyprus Research Promotion Foundation and the European Regional Development Fund through the project ΥΓΕΙΑ/ΔΥΓΕΙΑ/0308/11: Optimising Diagnostic Value in SPECT Myocardial Perfusion Imaging.

References
[1] Ficaro E P, Lee B C, Kritzman J N and Corbett J R 2007 *J. Nucl. Cardiol.* **14**(4) 455-465
[2] Germano G, Kavanagh P B, Slomka P J, Van Kriekinge S D, Pollard G and Berman D S 2007 *J. Nucl. Cardiol.* **14**(4) 433-454
[3] Garcia E V, Faber T L, Cooke C D, Folks R D, Chen J and Santana C 2007 *J. Nucl. Cardiol.* **14**(4) 420-432

Journal of Physics: Conference Series **317** (2011) 012020 doi:10.1088/1742-6596/317/1/012020

Fast analytic simulation toolkit for generation of 4D PET-MR data from real dynamic MR acquisitions

C Tsoumpas[1], C Buerger[1], P Mollet[2], P K Marsden[1]

[1] Division of Imaging Sciences & Biomedical Engineering, School of Medicine, King's College London, UK
[2] IBBT, University of Ghent, Belgium

E-mail: charalampos.tsoumpas@kcl.ac.uk

Abstract. This work introduces and evaluates a fast analytic simulation toolkit (FAST) for simulating dynamic PET-MR data from real MR acquisitions. Realistic radiotracer values are assigned to segmented MR images. PET data are generated using analytic forward-projections (including attenuation and Poisson statistics) with the reconstruction software STIR, which is also used to produce the PET images that are spatially and temporally correlated with the real MR images. The simulation is compared with the GATE Monte Carlo package, which has more accurate physical modelling but it is 150 times slower compared to FAST for ten respiratory positions and 7000× slower, when repeating the simulation. The region of interest for mean values and coefficients of variation obtained with FAST and GATE, from 65 million and 104 million coincidences, respectively, were compared. Agreement between the two different simulation methods is good. In particular, the percentage differences of the mean values are: 10% for liver, and 19% for the myocardium and a warm lesion. The utility of FAST is demonstrated with the simulation of multiple volunteers with different breathing patterns. The package will be used for studying the performance of reconstruction, motion correction and attenuation correction algorithms for dynamic simultaneous PET-MR data.

1. Introduction

The improvement of resolution and sensitivity of positron emission tomography (PET) has made apparent the need for motion correction. A critical element for the latter is the accurate representation of real motion vectors. In order to investigate the motion correction with realistic motion vectors, we devise a methodology for generation of simulated PET data with motion estimated from magnetic resonance (MR) images. A Fast Analytic Simulation Toolkit (FAST) has been developed to generate simulated simultaneous PET-MR datasets from acquired MR-derived radioactivity distributions with multiple respiratory positions for multiple breathing patterns and multiple volunteers.

2. Materials and methods

2.1. Simulation toolkit

We use the open-source software package STIR (release 2.1) [1] to generate PET data for the geometry of the Philips Gemini TF PET/CT scanner. The radioactivity (FDG) and attenuation distributions are assigned to segmented images. Multiple gated frames are generated for each respiratory position. The images are forward projected including attenuation effects to create sinograms.

2.2. Validation of simulation

PET projection data with 65 million and 104 million coincidences, respectively, were generated for one volunteer with FAST and validated against the same number of counts as generated with the Monte Carlo package GATE [2], which has more accurate physical modelling, but it is slower than the analytic method. For both methods, the same reconstruction and correction settings are used: OSEM (10 iterations, 23 subsets) and 3DRP reconstructions including corrections for attenuation and scatter [3] were performed with STIR.

2.3. Demonstration of simulation

A combination of static and dynamic MR acquisitions (35 frames × 0.7 s) is used to segment mainly the lungs from the tissue for multiple frames throughout the respiratory cycle [4]. Lesions of different sizes are inserted manually in one case. The segmented images, incorporated into FAST and PET images, are generated for five volunteers with different respiration pattern (normal and deep breathing).

3. Results

The results were validated against GATE for one respiratory position both visually (figure 1) and quantitatively in terms of regions of interest (ROI) mean values and coefficients of variation (COV) (Figure 2). Visual differences are noticeable in small regions due to the lack of detector modelling in analytic projections by FAST. Examples are demonstrated in figure 3 simulating multiple respiration positions from multiple volunteers with variable anatomy and breathing pattern. PET images correspond to five volunteers illustrated in descending size order from left to right. Two respiration patterns (normal and deep) have been acquired, showing one coronal slice at the end-inspiration and end-expiration for each case. All images relate spatially and temporally to the corresponding MR images demonstrating the fast generation of realistic simultaneous PET-MR dynamic datasets.

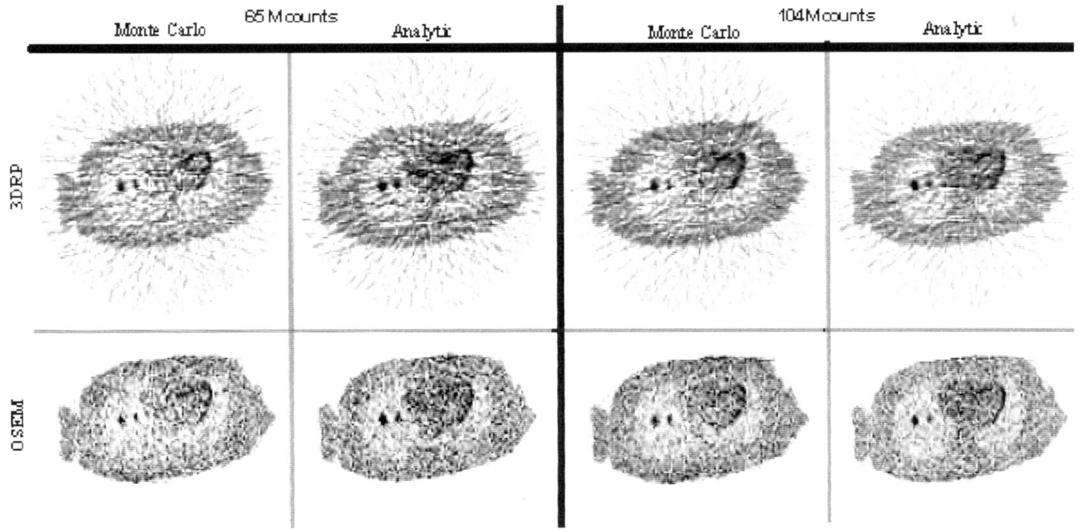

Figure 1. Analytic versus Monte Carlo simulations for data with two different noise levels (65 million counts and 104 million counts) reconstructed with 3DRP (upper row) and OSEM (lower row).

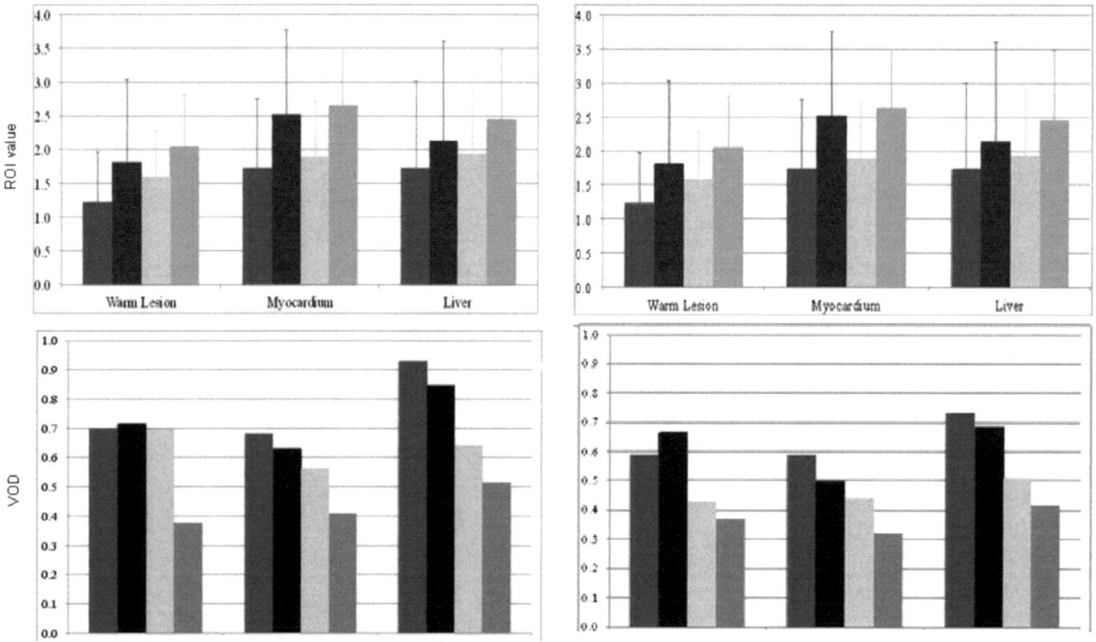

Figure 2. Upper row: mean ROI values for three regions (warm lesion, myocardium, liver) after reconstruction with OSEM and 3DRP for both GATE and analytic simulations; GATE-OSEM (dark green), Analytic-OSEM (dark blue), GATE-3DRP (light green), Analytic-3DRP (light blue). The error bars show the standard deviation of the mean ROI values. Lower row: COV values after reconstruction with OSEM and 3DRP for GATE and analytic simulations. Simulations are illustrated for two noise levels: 64 million counts (left column) and 104 million counts (right column).

4. Discussion

The analytic simulation method for two different noise levels was compared with GATE simulations at image level showing good agreement in terms of ROI and COV values for OSEM and 3DRP (figure 2). However, in figure 1 some visual difference is observed in small regions such as the three lesions or the myocardium, especially for 3DRP. This is attributed to the mismatch of detector response model in the two packages. The quantitative differences are, in 104 million counts, for OSEM liver 10%, OSEM myocardium and warm lesion (i.e. figure 1 - right side) 19%, while for 3DRP liver are 12%, 3DRP myocardium 17% and 3DRP warm lesion 12%. The COV, which is a measure of noise in the image, is similar for both simulations. In particular, the COV difference between the two methods is for OSEM liver 10%, OSEM myocardium 9%, and OSEM warm lesion 6%, while for 3DRP liver 10%, 3DRP myocardium 15% and 3DRP warm lesion 8%.

Figure 3 illustrates a sample of a PET-MR database of multiple simulated PET studies from real MR acquisitions, with different anatomy, and potentially different physiology. In this figure, two breathing patterns are shown (deep and normal), at two respiration positions for each cycle (end-inspiration, end-expiration). The advantage of using multiple dynamic MR images is that they provide a fast method for creating multiple frames of realistic simultaneous PET-MR data.

5. Conclusions

The importance of the results presented in this investigation is that analytic simulations can replace the time-consuming Monte Carlo simulations for emission tomography. The computational speedup factor is 150× from GATE to FAST and this is a clear advantage of this approach. Moreover, the speedup for repeating multiple realizations is 7000×. This provides the advantage of rapidly producing multiple simulations with different anatomical and physiological circumstances for any kind of clinical PET

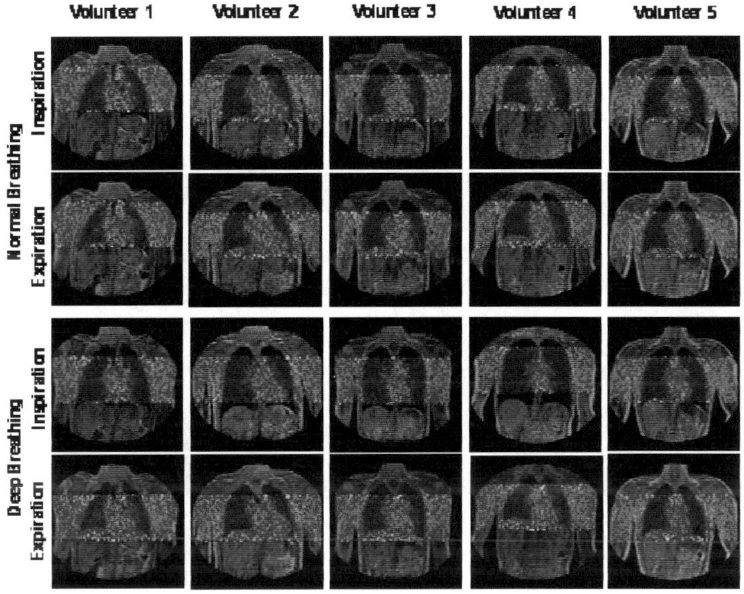

Figure 3. Multiple 3D dynamic PET-MR images with different respiration pattern: normal and deep breathing. In the figures, the two extreme positions of inspiration and expiration are shown. All images are displayed in a fused style with the same colour scale for PET, and grey scale for MR.

scanner. Although there is no claim that FAST can replace the accuracy of GATE or other Monte Carlo packages, it is expected that it will be useful for the evaluation of different reconstruction and correction strategies, especially for simultaneous PET-MR acquisitions.

Acknowledgements
This project is part of the HYPERimage project (http://www.hybrid-pet-mr.eu) supported by the European Union under the seventh framework program (No: 201651).

References
[1] Thielemans K, Mustafovic S and Tsoumpas C 2006 STIR: Software for Tomographic Image Reconstruction Release 2 *IEEE Nucl. Sci. Symp. Conf. Rec.* 4 pp 2174-76
[2] Jan S *et al* 2004 GATE: a simulation toolkit for PET and SPECT *Phys. Med. Biol.* **49** 4543-61
[3] Tsoumpas C, Aguiar P, Nikita K S, Ros D and Thielemans K 2004 Evaluation of the single scatter simulation algorithm implemented in the STIR library *IEEE Nucl. Sci. Symp. Conf. Rec.* pp 3361-5
[4] Tsoumpas C, Buerger C, King A P, Keereman V, Vandenberghe S, Schulz V, Schaeffter T and Marsden P K 2009 Simulation of dynamic PET data from real MR acquisitions *IEEE Nucl. Sci. Symp. and Med. Imag. Conf.* pp 3065-68

Journal of Physics: Conference Series **317** (2011) 012021 doi:10.1088/1742-6596/317/1/012021

Study of the effect of magnetic field in positron range using GATE simulation toolkit

G Soultanidis[1], N Karakatsanis[2], G Nikiforidis[1] and G Loudos[3]

[1] Department of Medical Physics, School of Medicine, University of Patras, Rion, Greece
[2] Biomedical Simulations and Imaging Applications Laboratory, School of Electrical and Computer Engineering, National Technical University of Athens, Greece
[3] Department of Medical Instruments Technology, Technological Educational Institute of Athens, Greece

E-mail: georgios.soultanidis@kcl.ac.uk

Abstract. In simultaneous PET-MR systems, the emitted positrons trajectory is influenced by the magnetic field. The aim of this study is to define that influence to the positron annihilation distribution. Monte Carlo methods have been applied, using GATE. Several isotopes were studied, in various types of materials and with different magnetic field strengths. The results showed variations in the positron range between different components and especially between the lung and water. Measurements of the 1-D positron annihilation distance indicated a reduction of the mean positron annihilation distance for ^{82}Rb of ~25%, ^{68}Ga of 19% and ^{18}F of 3.5%, at 3 Tesla. When the magnetic field was increased to 9.5 Tesla, the reduction was significant for all isotopes, and mainly for ^{68}Ga and ^{82}Rb, with approximately 41% reduction of the mean positron annihilation distance in water. Finally, the positron annihilation distribution varies according to the alignment with the magnetic field lines. The results of this study could be used to improve positron annihilation correction algorithms for simultaneous PET-MR acquisition, by taking under consideration the non-isotropic distribution.

1. Introduction

In the family of hybrid medical imaging systems the newest addition is the PET-MR. Using optic fibers or solid-state detectors, PET can be combined with MRI. As an example, RatCAP is the first system with the use of solid-state detectors [1]. Also, in PANDA and PANDA II systems, the light is guided outside the magnetic field and detected by PMTs [2, 3]. Finally, a simultaneous PET-MR system was developed by Judenhofer *et al.* for simultaneous scans with a 7 Tesla MRI [4]. It is well known that the magnetic field affects the path of a positron and therefore the final positron range. In fact, Raylman *et al.* describes this phenomenon with the use of Monte Carlo simulations. By his study, it is obvious that the magnetic field affects positron range and especially the equal distribution of positrons [5]. The main reason for studying positron range is to model it, compress range effects and make isotopes with higher positron emission energy suitable for PET imaging. Especially in small animal imaging, positron effect can be a strong limiting factor in overall resolution for isotopes such as ^{68}Ga and ^{82}Rb and hamper their application.

Bai *et al.* performed a study in order to correct the range effect for inhomogeneous materials, where EGS4 has been used [6]. Few studies on positron range are reported; as a complementary of the previous studies we used GATE, aiming to define the mean positron range for different materials and magnetic field amplitude, in order to present not only the 3-D positron range, but also the 1-D, along the axes lying parallel and perpendicularly to the magnetic field.

2. Materials and methods

In this study, the GATE (V 5.0.0_p01) simulation toolkit has been used [7]. GATE, as it is based on the Geant4 simulation toolkit, can simulate a vast background of physical processes. Preliminary studies of magnetic field effects have been carried out [8]. In GATE, the magnetic field can be adjusted in the three dimensional space. The influence of magnetic field upon each particle and the recalculation of its trajectory can be described by the Lorentz equation:

$$\vec{F}_{Lor} = q\vec{V}\times\vec{B} \qquad (1)$$

where q is the charge of the particle, B is the strength of the magnetic field and V is the velocity of the charged particle.

The materials used for this study were water, lung and ribs as defined in GATE. The isotopes selected for this study are: ^{18}F, ^{11}C, ^{15}O, ^{68}Ga and ^{82}Rb. Finally, the magnetic field strengths for this study are: 0, 1.5, 3 and 9.5 Tesla, selected in order to cover a wide range of clinical and experimental MR systems. The magnetic field was coordinated along the z-axis of the PET scanner.

The experiment geometry contained a single point source in the middle of a sphere with a radius of 40 mm. For each simulation ~ $20\text{x}10^6$ positron annihilations were measured. The software used for data acquisition and analysis was ROOT.

Data acquired in this study were the mean positron range in the 3-D plane and 1-D plane. The mean distance of positron travel is a (x, y, z) vector from the origin point to the annihilation point in the 3-D plane. The 1-D plane is the travel distance in a single dimension. The dimensions selected to be presented, are the x and z plane, perpendicular and parallel to the magnetic field, respectively.

3. Results

Table 1 presents the mean 3-D positron range while table 2 and table 3 give more specifically the mean 1-D range of positron at x and z planes, respectively.

Table 1. Mean 3-D Positron range (mm).

	Water				Lung				Rib Bone			
B_0	0 T	1.5 T	3 T	9.5 T	0 T	1.5 T	3 T	9.5 T	0 T	1.5 T	3 T	9.5 T
^{18}F	0.56	0.56	0.54	0.43	2.23	1.88	1.49	1.20	0.33	0.33	0.33	0.29
^{11}C	1.05	1.03	0.96	0.67	3.98	3.03	2.43	2.06	0.56	0.56	0.55	0.46
^{15}O	2.44	2.31	2.00	1.41	9.26	6.20	5.28	4.53	1.29	1.28	1.22	0.90
^{68}Ga	2.62	2.47	2.12	1.50	9.94	6.58	5.64	4.84	1.39	1.37	1.30	0.95
^{82}Rb	5.21	4.77	3.90	2.88	19.80	12.07	10.76	9.61	2.78	2.70	2.48	1.74

Table 2. Mean 1-D Positron range at x plane (mm).

	Water				Lung				Rib Bone			
B_0	0 T	1.5 T	3 T	9.5 T	0 T	1.5 T	3 T	9.5 T	0 T	1.5 T	3 T	9.5 T
^{18}F	0.27	0.27	0.26	0.16	0.73	0.65	0.46	0.19	0.16	0.16	0.16	0.13
^{11}C	0.44	0.43	0.40	0.21	1.21	0.96	0.61	0.25	0.26	0.26	0.25	0.19
^{15}O	0.90	0.86	0.73	0.32	2.66	1.68	0.99	0.34	0.53	0.52	0.50	0.31
^{68}Ga	0.93	0.89	0.76	0.33	2.74	1.75	1.03	0.35	0.56	0.55	0.52	0.32
^{82}Rb	1.87	1.68	1.27	0.50	6.07	2.79	1.60	0.48	1.05	1.02	0.93	0.48

Table 3. Mean 1-D Positron range at z plane (mm).

	Water				Lung				Rib Bone			
B_0	0 T	1.5 T	3 T	9.5 T	0 T	1.5 T	3 T	9.5 T	0 T	1.5 T	3 T	9.5 T
^{18}F	0.27	0.27	0.28	0.30	0.72	0.80	0.91	1.02	0.16	0.16	0.16	0.16
^{11}C	0.43	0.44	0.45	0.51	1.20	1.42	1.55	1.71	0.26	0.26	0.26	0.27
^{15}O	0.89	0.89	0.92	0.98	2.69	3.26	3.33	3.88	0.52	0.53	0.54	0.59
^{68}Ga	0.92	0.95	1.02	1.13	2.77	3.37	3.61	4.10	0.55	0.56	0.57	0.63
^{82}Rb	1.89	1.96	2.08	2.13	6.15	7.36	7.24	8.00	1.04	1.06	1.09	1.18

4. Discussion

In this study GATE was used to simulate the influence of the magnetic field to the positron range. The results were accurate, reliable and in agreement with the literature, where the mean positron range of fluorine in water at 0 Tesla is similar [6]. Yet, the literature is poor for other isotopes and their mean positron range. As it was expected by equation (1), isotopes with higher positron energy have the most spectacular reduction of mean 3-D positron range. The reduction of mean 3-D positron range is less for the broadly used fluorine but gives better statistics to other potential isotopes.

The variation of mean positron range between materials is enormous, especially in the case of the lungs. The big mean positron range indicates that, even if the reduction is big, for some isotopes a correction algorithm is necessary.

The mean 1-D positron range reveals another aspect of the influence of magnetic field. The mean 1-D positron range might be reduced in the x plane but the range in the z plane is increasing. This result brings new evidence about the need for re-evaluation of the existing positron range correction algorithms, in order to be adapted to PET applications inside magnetic fields.

To conclude, this study points out that GATE has the capability to model the presence of magnetic field and produce reliable results on positron range. The impact of the magnetic field on the spatial resolution can be modeled. Finally the data of anisotropic distribution of annihilation might be used as an input for improved positron range correction algorithms in PET-MR applications.

Acknowledgments
The first author would like to thank Professor Paul Marsden and Dr. Charalampos Tsoumpas for giving the permission to publish this study and for their support to his future plans.

References

[1] Woody C, *et al* 2007 Preliminary studies of a simultaneous PET/MRI scanner based on the RatCAP small animal tomograph *Nuclear Instruments & Methods in Physics Research Section A* **571** 102-5

[2] Mackewn J E, Strul D, Hallett W A, Halsted P, Page R A, Keevil S F, Williams S C R, Cherry S R and Marsden P K 2005 Design and development of an MR-compatible PET scanner for imaging small animals *IEEE Transactions on Nuclear Science* **52** 1376-80

[3] Marsden P K, Strul D, Keevil S F, Williams S C R and Cash D 2002 Simultaneous PET and NMR *British Journal of Radiology* **75** S53-S9

[4] Judenhofer M S, Catana C, Swann B K, Siegel S B, Jung W I, Nutt R E, Cherry S R, Claussen C D and Pichler B J 2007 PET/MR images acquired with a compact MR-compatiable PET detector in a 7-T magnet *Radiology* **244** 807-14

[5] Raylman R R, Hammer B E and Christensen N L 1996 Combined MRI-PET scanner: A Monte Carlo evaluation of the improvements in PET resolution due to the effects of a static homogeneous magnetic field *IEEE Transactions on Nuclear Science* **43** 2406-12

[6] Bai T, Ruangma A, Laforest R, Tai Y C and Leahy R M 2003 Positron range modeling for statistical PET image reconstruction *IEEE Nuclear Science Symposium Conference Record* **4** 2501-5

[7] Jan S, *et al* 2004 GATE: a simulation toolkit for PET and SPECT *Physics in Medicine and Biology* **49** 4543-61

[8] Jan S, Comtat C, Strul D, Santin G and Trebossen R 2005 Monte Carlo simulation for the ECAT EXACT HR plus system using GATE *IEEE Transactions on Nuclear Science* **52** 627-33

A comparative investigation of scatter correction in 3D PET

I Polycarpou, P K Marsden and C Tsoumpas

Division of Imaging Sciences & Biomedical Engineering, School of Medicine, King's College London, UK

E-mail: irene.polycarpou@kcl.ac.uk

Abstract. In 3D PET scatter degrades image quality and quantification. The currently most popular scatter estimation method is the single scatter simulation (SSS) which accommodates for multiple scattering by scaling the single scatter estimation. However, it has not been clear yet how accurate this approximation is for cases where multiple scatter is significant, raising the specific questions: "How important double scatter correction is, and how accurately do we simulate the total scatter events by appropriate scaling?" This project aims to clarify the improvements in terms of quantification due to scatter correction, using: (i) single scatter events only, (ii) single and double scatter events, (iii) total scatter events, or (iv) scaled single scatter, and evaluate the analytic scatter estimation algorithm as implemented in the open source reconstruction software STIR. The analytic SSS scatter estimation implemented in STIR is compared with the SimSET Monte Carlo package. Scatter correction accuracy is examined for different levels of scattering and scaling approaches. A large anthropomorphic phantom was reconstructed with FBP. The images have been compared quantitatively: Areas with high scatter fraction are compared with single scatter corrected images and show a 50% bias reduction after performing single and double scatter correction. Scaled single scatter correction results are in good agreement with SimSET true events, less than 10% difference. Total-fit and tail-fit scaled single scatter results in approximately equal mean values. SSS correction with tail-fit scaling in STIR is very close with SimSET true events, 10% difference. The results show that multiple scatter correction improves accuracy and scaling single scatter is an efficient method to compensate for multiple scattering for standard PET scanning acquisitions.

1. Introduction

In 3D PET scattered photons constitute 30-50% of the total measured data degrading not only image quality but also quantification. The currently most popular scatter estimation method is the single scatter simulation (SSS) [1, 2]. SSS uses emission and transmission images to estimate the scattered events along each line of response. The scatter distribution is calculated by summing the estimated scatter probabilities for each possible scatter point in the transmission image and each detector pair [3-4]. This algorithm was implemented within the open-source library software for tomographic image reconstruction (STIR release 2.1) [5], where the single scatter distribution is computed with SSS [6]. As suggested by Watson *et al* [7], the simplest approach to accommodate for multiple scatter is by scaling the scatter projection data for each plane to the tails of the measured data.

2. Methods

A larger version of an anthropomorphic numerical phantom was used [8]. PET data was simulated using the Simulation System for Emission Tomography (SimSET) [9] and the GE Discovery™ STE scanner. SimSET data included true, single scatter, double scatter and multiple scatter events.

STIR implementation of the three-dimensional filtered reprojection algorithm (3DRP) has been used. The x, y matrix size was 128×128 pixels with pixel size 5.27 mm, while along the z direction were 47 pixels with pixel size 3.27 mm. The cut-off frequency of the ramp filter was set to 0.5 cycles.

Scatter scaling was performed using the tail-fit or total-fit method. During tail fitting, the scale factor was estimated per slice so as to be equal with the counts in the tails of the emission (total non-scatter corrected) sinogram, where only scatter events are expected. To define this region, we use the attenuation correction factors sinogram with minimum and maximum threshold of 0.4 and 1.4, respectively. For total fit, one global scale factor was used by scaling the single scatter such that the total counts of single scatter are equal to the total scatter counts.

The different scatter orders were examined using the following sinograms (based on SimSET data):

- total (non-scatter corrected) - summation of the data of all events,
- single scatter corrected - summation of the data of all events apart from single scatter,
- single plus double scatter corrected - data of all events apart from single and double scatter,
- total scatter corrected - only the true (unscattered) events,
- total-scaled single scatter corrected - estimated by scaling the single scatter data using the total-fit and subtracting the result from the total data,
- tail-scaled single scatter corrected - estimated by scaling the single scatter data using the tail-fit and subtracting the result from the total data.

Finally, the scatter distribution was computed with the SSS algorithm implemented in STIR [6] using emission and transmission data together with the detector geometry. The emission data were the total (non-scatter corrected) SimSET data. The scatter estimation was applied to the emission data iteratively to perform scatter correction [2]. Firstly, the emission data are corrected for attenuation and reconstructed for input to SSS. The SSS sinogram was "subsampled" by using the same scanner dimensions but with less number of rings and detectors. After upsampling the SSS, it is subtracted from the emission data to get a scatter-corrected sinogram. Then the SSS is repeated using as input the scatter-corrected reconstructed sinogram. Finally, the two scatter estimates are averaged and the result is used for scatter correction.

3. Results

Figure 1 illustrates the coronal planes for the 3DRP reconstructed images of SimSET data for total (non-scatter corrected), single scatter corrected, single plus double scatter corrected, total scatter corrected, total-scaled single scatter corrected and tail-scaled single scatter corrected data. Figure 2 shows plotted profiles through the same transaxial plane of the images of SimSET data for true events, single scatter corrected and single plus double scatter corrected or for true events, total-scaled single scatter corrected and tail-scaled single scatter corrected.

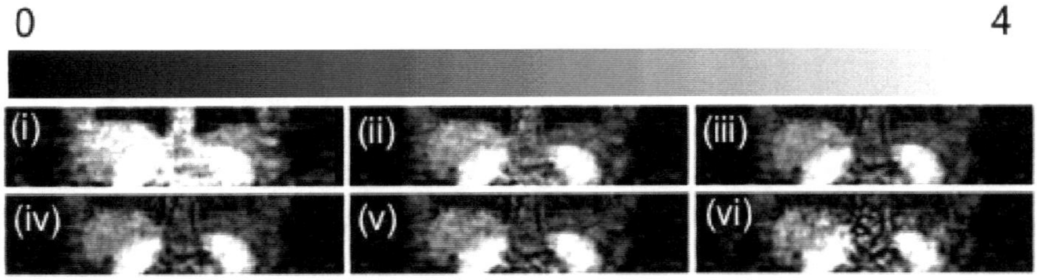

Figure 1. SimSET data: (i) total, (ii) single scatter corrected, (iii) single plus double scatter corrected, (iv) total scatter corrected, (v) total-scaled single scatter corrected, (vi) tail-scaled single scatter corrected.

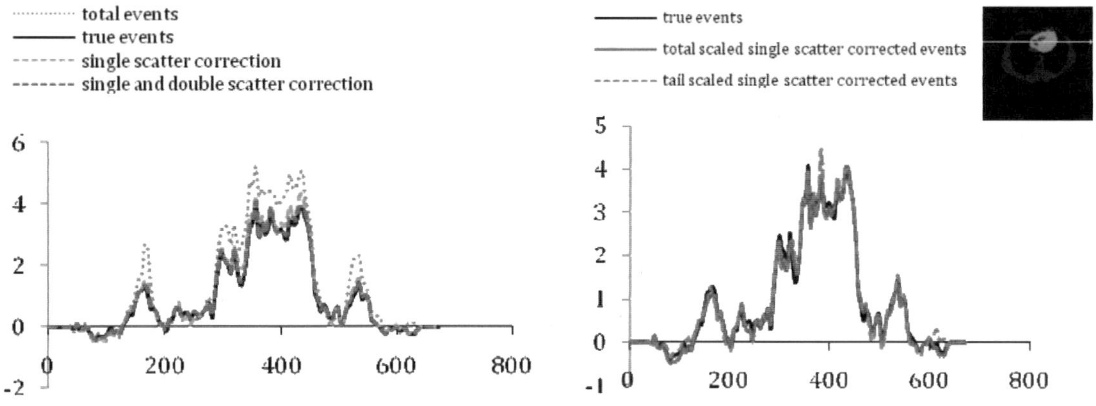

Figure 2. Image (top right) profiles of different investigated cases.

Figure 3 presents transverse slices of 3DRP reconstructed images for the scatter corrected data with STIR and SimSET unscattered events.

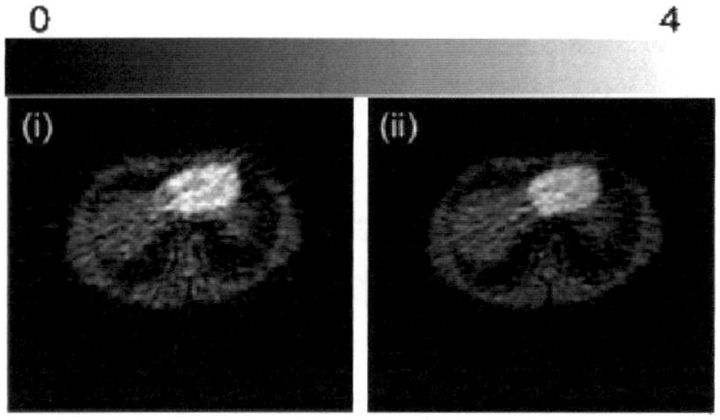

Figure 3. Images of (i) SimSET scatter corrected data with STIR, and (ii) SimSET total scatter corrected.

Figure 4 illustrates histogram of the ROI mean values of the ideal distribution and all the investigated cases.

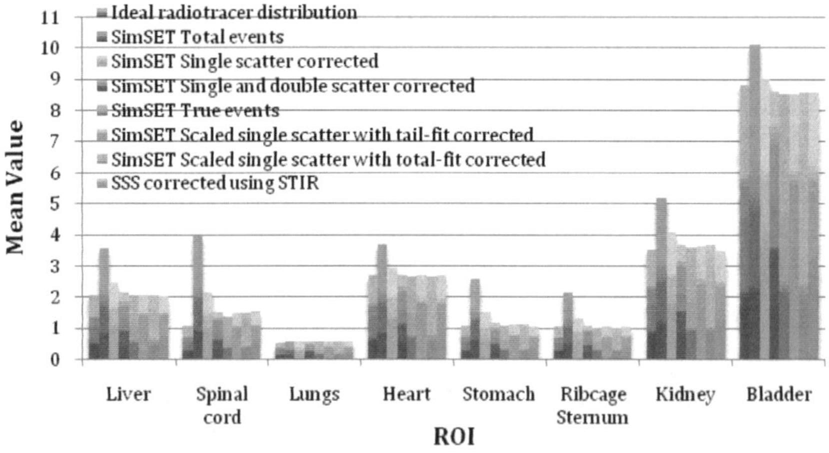

Figure 4. Histogram of ROI mean values of the ideal distribution and all the investigated cases.

4. Discussion

Figures 1 and 4 indicate that scatter affects quantitative accuracy and show visual and quantitative differences after performing double scatter correction. In figure 4, for areas with high scatter fraction, comparing single scatter correction and single plus double scatter correction, there is important mean value reduction after performing single plus double scatter correction.

Previous studies showed that multiple scattering broadens the scatter distribution and thus single and total scatter distribution can have different shapes depending on the object attenuation and the detector settings [10]. In our simulations, figure 1 shows that the image of single plus double scatter corrected has similar appearance to the single scatter corrected. Figures 1 and 2 show that the tail- or total-fit scaled single scatter corrected images is in a good approximation with the free of scatter image. In addition, figures 2-4 verify that scaled single scatter correction is in good agreement with SimSET true events with a difference no more than 10%. Total-fit and tail-fit scaled single scatter results have approximately equal mean values.

Finally, in figure 3 and 4, SSS correction with tail-fit scaling in STIR has very close visual and quantitative results with SimSET true events, 10% difference.

5. Conclusions

Multiple scatter correction appears to improve image contrast and accuracy, and therefore it is essential in PET imaging for precise estimation of activity distribution. The results show that for standard PET scanning acquisitions, scaling single scatter is an efficient method to compensate for multiple scattering. Finally, SSS correction with tail-fit scaling in STIR is in good agreement with SimSET simulations.

Acknowledgments

This investigation was for the partial fulfilment of the M.Sc. degree of I. Polycarpou at King's College London, accomplished at St Thomas Hospital, London. Special thanks go to Dr. Kris Thielemans and Dr. Ravindra Manjeshwar (GE Healthcare). We would also like to thank Dr. Slavik Tabakov.

References

[1] Watson C, Newport D and Casey M 1996 A single scatter simulation technique for scatter correction in 3D PET *Three-Dim. Imag. Recon. Rad. Nucl. Med.* **4** 255–68

[2] Watson C, Newport D, Casey M, DeKemp R, Beanlands R and Schmand M 1997 Evaluation of simulation-based scatter correction for 3-D PET cardiac imaging *IEEE Nucl. Sci. Symp. Conf. Rec.* **44** 90-97

[3] Zaidi H and Koral K 2004 Scatter modelling and compensation in emission tomography *Eur. J. Nucl. Med. Mol. Imaging* **31** 761-82

[4] Werling A 2002 Model-based scatter correction for positron emission tomography (in German) *Med. Phys.* **29** 105

[5] Thielemans K, Mustafovic S and Tsoumpas C 2007 STIR: Software for tomographic image reconstruction Release 2 *IEEE Nucl. Sci. Symp. Med. Imag. Conf. Rec.* 2174-76

[6] Tsoumpas C, Aguiar P, Nikita K S, Ros D and Thielemans K 2004 Evaluation of the single scatter simulation algorithm implemented in the STIR library *IEEE Nucl. Sci. Symp. Med. Imag. Conf. Rec.* **6** 3361-65

[7] Watson C 2000 New, faster, image-based scatter correction for 3D PET *IEEE Nucl. Sci. Symp. Med. Imag. Conf. Rec.* **47** 1587-94

[8] Zubal I G, Harrell C R, Smith E O, Rattner Z, Gindi G and Hoffer P B 1994 Computerized three-dimensional segmented human anatomy *Med. Phys.* **21** 299-302

[9] Lewellen T, Harrison R and Vannoy S 1998 The SIMSET program in Monte Carlo calculations in nuclear medicine *IOP Publishers* 77-92

[10] Tsoumpas C, Aguiar P, Ros D, Dikaios N and Thielemans K 2005 Scatter simulation including double scatter *IEEE Nucl. Sci. Symp. Med. Imag. Conf.* **3** 1615-19

Journal of Physics: Conference Series **317** (2011) 012023 doi:10.1088/1742-6596/317/1/012023

NEMA NU-1 2007 based and independent quality control software for gamma cameras and SPECT

A Vickery[1], T Jørgensen[2] and R de Nijs[3]

[1] Department of Clinical Physiology and Nuclear Medicine, Glostrup Hospital, Denmark
[2] Department of Clinical Physiology and Nuclear Medicine, Naestved Hospital, Denmark
[3] Department of Clinical Physiology, Nuclear Medicine & PET, Rigshospitalet, Copenhagen University Hospital, Denmark

E-mail: anette@vickery.dk

Abstract. A thorough quality assurance of gamma and SPECT cameras requires a careful handling of the measured quality control (QC) data. Most gamma camera manufacturers provide the users with camera specific QC Software. This QC software is indeed a useful tool for the following of day-to-day performance of a single camera. However, when it comes to objective performance comparison of different gamma cameras and a deeper understanding of the calculated numbers, the use of camera specific QC software without access to the source code is rather avoided. Calculations and definitions might differ, and manufacturer independent standardized results are preferred. Based upon the NEMA Standards Publication NU 1-2007, we have developed a suite of easy-to-use data handling software for processing acquired QC data providing the user with instructive images and text files with the results.

1. Introduction

Within Nuclear Medicine, image abnormalities and artifacts affecting the quality of images are well known phenomena [1]. Therefore, it is of great importance to have a thorough Quality Assurance for gamma and SPECT cameras to minimize the occurrence of these abnormalities and artifacts. EANM has made recommendations of routine quality control for nuclear medicine instrumentation [2]. After installation and before the camera is put into clinical use, it should undergo National Electrical Manufacturers Association (NEMA) Performance standard measurements to verify that the camera performs according to specification supplied by the manufacturer and to establish baseline conditions for all future measurements [3]. The NEMA Standards Publication NU 1-2007 describes how to perform, process and report QC tests for gamma and SPECT cameras [3]. Often, with support from the manufacturers, all necessary phantoms can be supplied and acquisitions can be done according to NEMA, but a thorough Quality Assurance also requires a careful handling of the measured QC data.

The lack of a manufacturer independent QC-software supporting a NEMA performance standard is a major problem performing NEMA QC tests. NEMA based software exists [4], however the number of tests is limited. Therefore, we have developed a full suite of data handling software based upon the NEMA Standard Publication NU-1 2007. Our QC software complements camera specific manufacturers' software by providing an independent processing platform for different cameras.

2. Methods and design

The software is based on NEMA recommendations regarding processing and analysis of the data [3], and runs in MATLAB (Mathworks, Natick, Massachusetts, USA). It is capable of calculating Intrinsic Spatial Resolution and Linearity (ISR & ISL), Intrinsic Energy Resolution (IER), Intrinsic Flood Field Uniformity (IFFU), Multiple Window Spatial Resolution (MWSR), Intrinsic Count Rate Performance (ICRP), System Spatial/Scan Resolution (SSPR & SSCR), System Planar Sensitivity (SPS), System Alignment (SA), System Volume Sensitivity (SVS) and SPECT Reconstructed Spatial Resolution (SRSR).

The data handling programs are aimed at making the processing of acquired QC-data a simple task, providing the users with instructive images of the processed data as well as text files containing the results of the QC.

3. Example: Intrinsic spatial resolution and linearity

The image required for the calculation of Intrinsic Spatial Resolution and Linearity is acquired using a special lead mask (3 mm thick for Tc-99m) covering the entire Field Of View (FOV) with 1 mm wide parallel slits. The FOV is exposed by a Tc-99m point source positioned at distance of at least 5 times the maximum dimension of Useful FOV (UFOV) [3]. Figure 1 (left) shows an example of an acquired image for the determination of the ISR&L in the Y direction (trans-axial). The frames indicate the UFOV and Camera FOV (CFOV). By summing the image parallel to the direction of the slits, a number of line-spread functions is generated. Each line-spread function is characterized by a number of peaks corresponding to the number of visible slits. The main task of the Intrinsic Spatial Resolution analysis is to determine the location and the Full Width at Half Maximum (FWHM) and the Full Width at Tenth Maximum (FWTM) of these peaks.

The intrinsic spatial differential linearity is calculated as the standard deviation of the peak-locations in each slit. The intrinsic spatial absolute linearity is determined by fitting a set of equally spaced parallel lines to the peak-locations. The intrinsic spatial absolute linearity is then determined as the maximum deviation of the peak locations from the grid fit. Only if the lead mask is perfectly aligned, the grid will be orthogonal. Figure 1 (right) shows the difference between the peak locations and the best-fit orthogonal grid. Towards the left, the difference is positive, while it is negative towards the right. This indicates that the lead mask may be rotated by a small angle. By repeating the analysis, it turns out that the optimal fit is obtained by rotating the lines by 0.09 degrees, and this improves the result for the intrinsic spatial absolute linearity from 0.66 mm to 0.31 mm. Results from the processed images are shown in table 2.

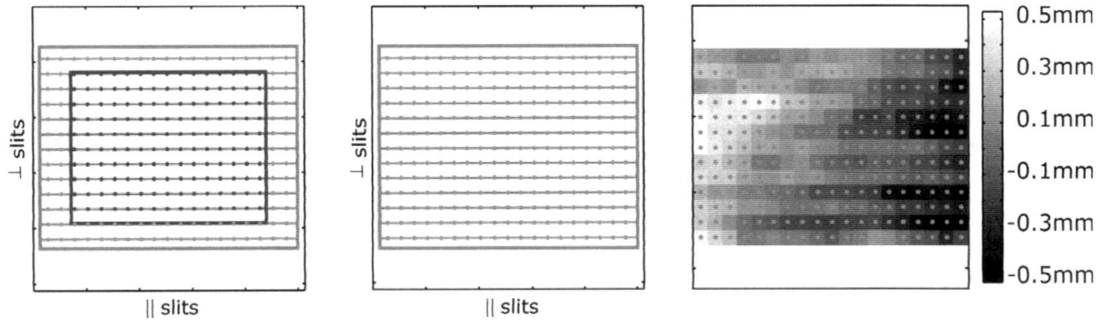

Figure 1. Image acquired for the determination of the ISR&L in the Y direction (trans-axial). The frames indicate the UFOV and CFOV. The peak locations are superimposed (left). The best orthogonal grid fit superimposed on the image. The frame indicates the UFOV (middle). The difference between the peak locations and the orthogonal grid (right).

Journal of Physics: Conference Series **317** (2011) 012023 doi:10.1088/1742-6596/317/1/012023

4. A novel method: Resolution and pixel size using frequency analysis

For daily qualitative inspection, flat lead or tungsten phantoms are often used in combination with a Co-57 flood source. However, a quantitative analysis of these measurements is preferred. For determining pixel size and resolution at 0 cm from the collimator, a 3 mm thick orthogonal hole lead phantom with 4 mm holes at 4 mm distance, PTW Freiburg L991151 QUASI 3, and a bar phantom with four resolution quadrants (2.5, 3, 3.5 and 4 mm slits with identical spacing) were used. The phantoms were directly placed on the collimator surface with a Co-57 flood source on top. Two to four million counts in a 256x256 matrix with 1.066 mm pixel size were acquired per image in a 122 keV (20% width) energy window. No filtering or resampling was performed.

In the discrete fourier transform, the sampling distance in frequency space, Δf in cm^{-1} and FOV in cm, are related reciprocally by

$$\Delta f = \text{FOV}^{-1} = (N \cdot d)^{-1}, \tag{1}$$

with N the number of samples and d the pixel size in cm. This gives the possibility to relate a position i_f, the number of samples times the sampling distance, in frequency space to pixel size by

$$f = i_f \cdot \Delta f = (N \cdot d)^{-1}. \tag{2}$$

If the discrete fourier transform is applied to the images acquired with an orthogonal hole or bar phantom, this equation can be used for measuring the (average) pixel size in two directions. A non-integer value for i_f is determined by an intensity-weighted determination of the position in frequency space.

If a gaussian point spread function is assumed, it is possible to relate the intensities in fourier space to the FWHM of the point spread function. For a gaussian, the standard deviations σ in cartesian denoted with subscript x, and frequency space denoted with subscript f, are [5] related by

$$\sigma_x \cdot \sigma_f = c\,(2\pi)^{-1}, \tag{3}$$

while FWHM and σ are related by

$$\text{FWHM}^2 = 8 \ln 2 \cdot \sigma^2. \tag{4}$$

The FWHM in cartesian space is calculated from σ in frequency space.

An example (data from a Philips Precedence SPECT/CT) of the bar phantom and its 2D fourier transform is shown in figure 2. The resulting frequency is given as the reciprocal value of the dominant wavelength, which is 8 mm in the case of 4 mm holes/lines with identical spacing. The resolution is obtained by fitting the four intensities at the frequencies corresponding to the line distance with a gauss curve or linearly, in figure 3, by plotting the natural logarithm of the intensity against the squared frequency.

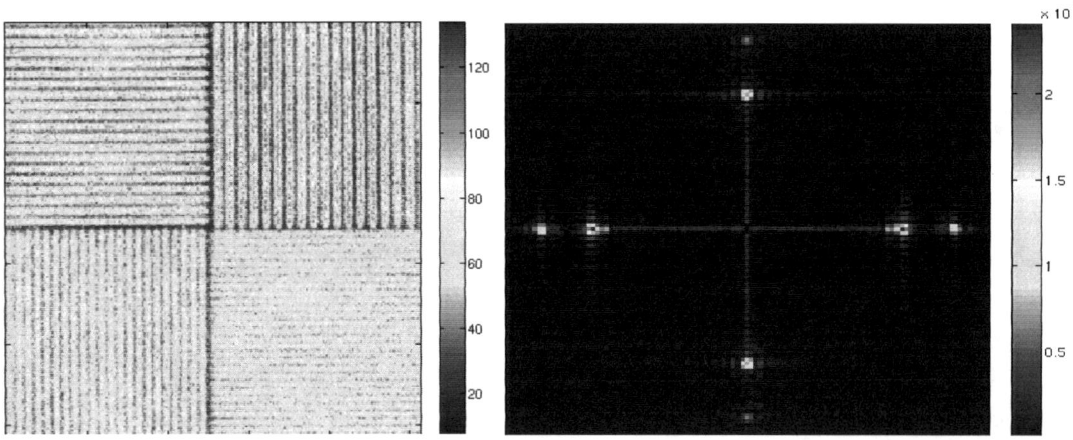

Figure 2. Acquired image of the bar phantom (left) and its zoomed 2D fourier transform (right). The DC component in the center is set to zero.

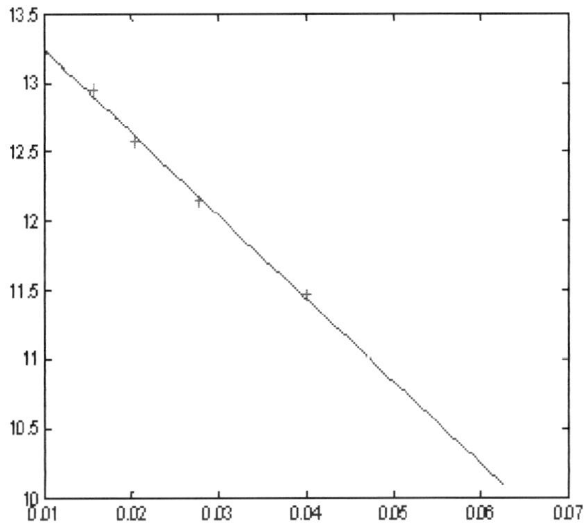

Figure 3. The natural logarithm of the intensity is plotted against the squared frequency in mm^{-2} and fitted linearly. The slope is a measure for the resolution.

The obtained values were compared with NEMA measurements for two line sources of Tc-99m, placed directly on the collimator. The results are shown in table 1. The pixel size values determined with the phantoms correspond to the scanner specifications and the NEMA line source measurements. The resolution is close to the resolution determined by the line source measurements. The measured resolution can be considered as a measurement of the intrinsic resolution, if the hole diameter is small, since the phantom is placed at zero distance. The hole size, which normally is 1-1.5 mm for low energy collimators, contributes to the measured resolution with typically a few tenths of a mm. With this way, it might be more difficult to detect errors in the collimator.

Table 1. Comparison of the fourier method with NEMA-measurements. Pixel size and (planar) resolution are averaged in both directions.

	Line source	Bar phantom	Hole phantom	Specifications
Pixel size (mm)	1.065	1.066	1.071	1.066
Resolution (mm)	3.93	3.83, 3.95 [a]	-	4.0

[a] Gaussian, linear fit

Alternatively, the phantoms and flood source can be positioned at 10 cm distance, but this is not always easily achievable in clinical practice.

5. Results

Results processed with different software and for different SPECT are shown in table 2 and 3. Table 2 shows the output for ISR and ISL. A selection of calculated NEMA-specification is shown in table 3. Due to the manufacturers' software, that only supports calculation of some NEMA specifications, it was not possible to process several NEMA parameters indicated by a "-" in table 3, for the SymbiaT16 and Discovery.

6. Discussion and conclusions

We have demonstrated that our software is able to calculate several NEMA-specifications, and where possible, the results were compared to those obtained by the manufacturers. As shown in table 2 and 3, there is an agreement between our software and the manufacturers' software. A novel quantitative analysis of orthogonal hole and quadrant phantoms was presented.

Table 2. ISR and ISL results based upon NEMA Standard Publication NU1- 2007 processed with manufactures software and our software. Results are shown for x-direction for detector 1.

	SymbiaT16 manuf. results			SymbiaT16 our results			Discovery manuf. results (Xeleris)			Discovery our results	
ISR-x	FWTM (mm)	FWHM (mm)	**ISR-x**	FWTM (mm)	FWHM (mm)	**ISR-x**	FWTM (mm)	FWHM (mm)	**ISR-x**	FWTM (mm)	FWHM (mm)
CFOV	6.9	3.6	CFOV	6.9	3.7	CFOV	6.5	3.5	CFOV	6.6	3.5
UFOV	7.1	3.7	UFOV	7.1	3.7	UFOV	6.6	3.5	UFOV	6.6	3.5
ISL-x	Diff. lin. (mm)	Abs. lin. (mm)	**ISL-x**	Diff. lin. (mm)	Abs. lin. (mm)	**ISL-x**	Diff. lin. (mm)	Abs. lin. (mm)	**ISL-x**	Diff. lin. (mm)	Abs. lin. (mm)
CFOV	0.1	0.2	CFOV	0.1	0.2	CFOV	0.0	0.3	CFOV	0.2	0.3
UFOV	0.1	0.4	UFOV	0.1	0.3	UFOV	0.1	0.3	UFOV	0.2	0.3

Table 3. A small selection of results based upon NEMA Standard Publication NU1- 2007 calculated by different software.

	SymbiaT16 manuf. results	SymbiaT16 our results	Discovery manuf. results	Discovery our results
SPS	LEHR coll.	LEHR coll.	LEHR coll.	LEHR coll.
(Cps/MBq)	-	92	-	74
IER				
FWHM (%)	8.6	8.9	-	-
SSPR	LEHR coll.	LEHR coll.	LEHR coll.	LEHR coll.
FWHM (mm)	-	7.5	7.6	7.6
IFFU				
UFOVI (%)	2.1	2.5	2.4	2.2
CFOVI (%)	2.0	2.3	2.1	2.2
SA				
Cor. 1 (mm)	-	7.3	-	1.7
Cor. 12 (mm)	-	1.02	-	0.6
Axial 1 (mm)	-	-	-	0.5
Axial 12 (mm)	-	0.77	-	0.5

The selection of the platform used to code the software was MATLAB, which, in principal, is independent of the operating system. Future development could be the compilation of our software, making it independent of the MATLAB environment and an automatic report-generation using e.g. LaTeX. It is our vision that the software will be open access.

References

[1] International Atomic Energy Agency 2003 *IAEA quality control atlas for scintillation camera systems*

[2] Sokole E B, Płachcínska A and Britten A 2010 Routine quality control recommendations for nuclear medicine instrumentation *Eur. J. Nucl. Med. Mol. Imaging* **37** 662–671

[3] National Electrical Manufacturers Association *Performance measurements of gamma cameras* NEMA Standard Publication NU 1-2007

[4] Rova A, Celler A and Hamarneh G 2008 Development of NEMA-based software for gamma camera quality control *J. Digital Imaging* **21** 243-255

[5] Weisstein E W *Fourier Transform - Gaussian* From MathWorld - A Wolfram Web Resource http://mathworld.wolfram.com/FourierTransformGaussian.html

Specificity and sensitivity of SPECT myocardial perfusion studies at the Nuclear Medicine Department of the Limassol General Hospital in Cyprus

S Koumna[1,2], Ch Yiannakkaras[3], P Avraamides[2] and O Demetriadou[4]

[1] Athens Anticancer – Oncology Hospital "Agios Savvas", Athens, Greece
[2] Cardiology Clinic, Limassol General Hospital, Limassol, Cyprus
[3] Medical Physics Department, Nicosia General Hospital, Nicosia, Cyprus
[4] Nuclear Medicine Department, Limassol General Hospital, Limassol, Cyprus

Email: stelkoum@gmail.com

Abstract. The aim is to determine the sensitivity and specificity of Myocardial Perfusion Imaging (MPI) performed at the Nuclear Medicine Department of the Limassol General Hospital in Cyprus. Through a retrospective analysis, patient results obtained by MPI were compared to results obtained by Invasive Angiography. We analyzed data from 96 patients that underwent both MPI and Angiography during the years 2009-2010, with a maximum time interval of ± 9 months between the two types of medical exams. For 51 patients, the indication was the detection of CAD. For 45 patients, the indication was to assess viability and/or ischemia after MI, PCI or CABG. Out of 84 patients with CAD confirmed by angiography, 80 patients resulted in abnormal MPI (sensitivity of 95% and positive predictive value of 98%). Out of 12 patients with normal coronaries, 10 patients resulted in normal MPI (specificity of 83% and negative predictive value of 71%).In conclusion, for the patients with abnormal MPI and confirmed CAD, MPI was a useful aid for further therapy management.

1. Introduction

Myocardial Perfusion Imaging (MPI) has proven its value as a noninvasive examination, which determines whether or not a patient suffers from CAD [1-3]. When a patient is complaining of chest pain or other equivocal symptoms and when a stenosis is found with angiography, it is important to determine the clinical significance of the stenosis, to allow the cardiologist to proceed to further therapeutic intervention. The best way to estimate the efficiency of blood flow to a certain area and simultaneously to observe and calculate the function of the Left Ventricle is Gated SPECT MPI. This study aims at determining the sensitivity and specificity of MPI at the Nuclear Medicine Department of the Limassol General Hospital in Cyprus.

2. Methods

This is a retrospective analysis of the data obtained in 2009-2010 by MPI and Angiography in Limassol General Hospital Nuclear Medicine and Angiography Departments. 96 patients have been selected (78 male, 18 female) that fulfilled a requirement for a maximum of nine month interval between the two medical tests. The MPI was performed using one-day protocol with $Tc\text{-}99^m$ – Tetrofosmin. The type of stress was either dynamic or pharmaceutical [4].

The indications for MPI were separated into two categories. The first category referred to the detection of CAD (51 patients) whereas the second category referred to the assessment of viability and/or ischemia after MI, PCI or CABG (45 patients).

The gender of the patients was also considered in the analysis, as it affects the imaging results depending on soft tissue attenuation due to the diaphragm in men and due to breasts in women. Furthermore, we anatomically matched the defects detected with MPI against the findings of angiography. The stenosis type was categorized according to the stenosis of the vessel diameter: moderate stenosis (40-70%), significant stenosis (71-90%), tight stenosis (91-99%), occluded (100%).

3. Results

Among the 84 patients with CAD confirmed by angiography, 80 presented abnormal MPI test (sensitivity 95%, table 1 and 2), and 4 of them presented a normal MPI test (table 3). 12 patients did not present CAD confirmed by angiography, 10 of which had a normal MPI result (specificity 83%) (table 1 and 2).

Table 1. Number of patients with positive and negative MPI and angiography for CAD, respectively.

Vessel condition [a]	MPI	Angiography
CAD	80	84
No active CAD [a]	10	12

[a] Coronaries with good previous PCI result and no intrastent stenosis were considered as normal.

Table 2. Statistical analysis.

Statistical Parameter	Value [a] (%)	95% Confidence Interval (%)
Sensitivity	95	88-98
Specificity	83	51-97
Positive Predictive Value	98	91-100
Negative Predictive Value	71	42-90

[a] Approximate integer number.

Table 3. Patients with false negative result vessel condition.

Patient	Condition
1	LAD (patient PCI), D1 significant stenosis, LCx moderate stenosis, OM tight stenosis, RCA moderate stenosis
2	LMS 50% and LAD tight stenosis
3	LMS 90% and OM tight stenosis
4	Moderate RCA stenosis and distal LAD occlusion

For the 4 patients with false negative result, 1 of them presented three vessel disease, 2 presented LMS disease and 1 presented moderate RCA stenosis and distal LAD occlusion. Further analysis with anatomical matching of areas of defect with vessel territories, indicated that out of the 80 patients with true positive result, 8 (10%) were underestimated by MPI and 3 (4 %) were not matched [5]. From the 8 patients that were underestimated (table 4), 3 presented three-vessel disease, 3 presented three-vessel disease and LMS disease, 1 presented four-vessel disease and 1 presented one-vessel disease and LMS disease.

As for the gender consideration, false negative (FN) results have been recorded for 4 male patients, whereas false positive (FP) results have been observed for 1 male and 1 female patient. True positive results that were underestimated by MPI, have been recorded for 7 male patients and 1 female patient. Results that did not match have been observed for 3 male patients. The FP results have indicated a female patient with anterior wall defects and a male patient with inferior and anterior wall defects.

The obtained sensitivity and specificity (see table 2) of MPI for the detection and assessment of ischemia, in correlation within evasive angiography, are satisfactory when compared to published results (pooled sensitivity of 87% and specificity of 73% with exercise stress for 33 published studies, and pooled sensitivity of 89% and specificity of 75% with pharmacological stress for 17 published studies) [6-7].

Table 4. Condition of vessels that were underestimated with detected MPI defects.

Patient	Vessel Disease	Segments with defect
1	Three-vessel disease	4, 5, 11, 2, 3, 8, 9, 14
2	Three-vessel disease	4, 1, 7, 13
3	One-vessel disease and moderate LMS stenosis	4, 10, 15
4	Three-vessel disease and moderate LMS stenosis	5, 11, 17
5	Three-vessel disease and significant LMS stenosis	16
6	Three- vessel disease and significant LMS stenosis	4, 10, 15
7	Three-vessel disease	11, 5, 4
8	Four-vessel disease	8, 14, 10, 15

4. Discussion

These results should be seen under the perspective that the majority of the patients had a high pretest probability for CAD [8]. It is widely understood that when MPI presents a False Negative (FN) result the patient will not undergo angiography, and that not all FN will be detected. On the contrary, when a FP result is presented the patient will most probably undergo angiography, so the majority of FP results are identified. With this way, the sensitivity and specificity percentages are affected [9-10]. In this study the number of FN or FP results was not sufficient to enable further statistical analysis.
The FN detected results indicated LMS disease and three vessel disease in the majority of patients. This finding brings forward the necessity of Gated SPECT MPI so that in the absence of perfusion defects, a poor LV function, wall motion abnormality or other extra cardiac findings such as lung uptake to enable identification of multi-vessel CAD [11,12].

Because of the well known imaging problems with breasts and diaphragm attenuation in female and male patients, respectively, even after application of attenuation correction with gadolinium sources, we decided to utilize the 2-day protocol in obese patients so that to minimize the stress-rest photon overlapping in the 1-day protocol, which is mistakenly interpreted as reversible ischemia. The gender analysis showed a tendency for false/underestimating results in male patients. However, the total number of samples was insufficient for safer assumptions.

5. Conclusion

MPI was a threshold point for further therapeutic management for the patients that were correctly diagnosed with CAD or for the follow up of the patients after MI, PCI or CABG. In addition, MPI is useful for risk stratification when perfusion defects, function abnormalities and lung uptake are co-estimated [13-14]. It is though necessary to undertake a new prospective study with a sufficient normal group of patients in order to estimate sensitivity and specificity without referral bias.

References

[1] Machecourt J and Vanzetto G 2001 *Eur. Heart J. Suppl.* **3** F2-F4
[2] Stirrup J *et al* 2009 *Br. Med. Bull.* **89** 63-78
[3] Fard-Esfahani A *et al* 2009 *Hellenic J. Cardiol.* **50** 396-401
[4] Navare *et al* 2004 *J. Nucl. Cardiol.* **11** 551-61
[5] Rasulova N *et al* 2008 *Nucl. Med. Commun.* **29** 129-36
[6] Zaman M M *et al* 2010 *Mymensingh Med. J.* **19** 608-13
[7] Fuster V *et al* 2004 *Hurst's the Heart* 11th edition McGraw Hill Companies, Inc.
[8] Hendel R *et al* 2009 *J. Am. Coll. Cardiol.* **53** 2201–29
[9] Klocke *et al* 2003 *Circulation* **108** 1404-18
[10] Loong C Y and Anagnostopoulos C 2004 *Heart Suppl.* **90** V2-9
[11] Reyes E 2010 *Euro. Intervention Suppl.* **G** *72-8*
[12] Abidov *et al* 2003 *J. Am. Coll. Cardiol.* **42** 1818–25
[13] Mowatt G *et al* 2005 *Nucl. Med. Commun.* **26** 217-29
[14] Pavin D *et al* 1997 *European Heart Journal* **18** 69-77

Journal of Physics: Conference Series **317** (2011) 012025 doi:10.1088/1742-6596/317/1/012025

The activities of EFOMP in education and training: past, present and future

S Christofides

EFOMP President, Medical Physics Department, Nicosia General Hospital, Nicosia, Cyprus

E-mail: cstelios@cytanet.com.cy

Abstract. One of the main aims of the European Federation of Organisations for Medical Physics is to propose guidelines for education, training and accreditation programmes. This is achieved through the publication of Policy Statements and the organisation of education and training courses, seminars and conferences. It represents a long-term work-programme aimed at harmonising the education and training of the Medical Physicist across Europe. This paper presents an overview of the past, present and future efforts of EFOMP to achieve this aim.

1. Introduction

The European Federation for Organisations of Medical Physics - EFOMP was set up in 1980 in London (www.efomp.org). Among its aims and objectives is the production of guidelines for education, training and accreditation programmes.

EFOMP's first Policy Statement [1] was on "Medical Physics Education and Training: The present European Level and Recommendations for its Future Development", published in 1984. This policy statement is now obsolete and has been replaced by policy statement No. 12 [2].

An essential component of professionalism is Continuous Professional Development - CPD. EFOMP has been, and continues to be, very active in this area by publishing policy statements for CPD schemes and establishing National Accreditation systems for Medical Physicists.

The exchange of information is a very important component of CPD and for the education and training of any professional. Therefore the support, sponsorship and organisation of educational as well as scientific meetings is another essential component of the work that EFOMP is carrying out to advance the professionalism of the Medical Physicist in Europe.

Presently, EFOMP is an active partner in a number of European and International projects with the aim of producing guideline documents for the education and training of the Medical Physicists and other Healthcare Professionals using ionising radiation.

The purpose of this paper is to give an outline of the EFOMP policy statements its present activities and its future plans.

2. EFOMP Education and Training Activities

2.1. Policy Statements

In total, EFOMP has published 13 policy statements. All of them are related to the professional status, education, training, competence, roles and responsibilities of the Medical Physicist in Europe. A brief description of them is given below:

- Policy Statement 1: "Medical Physics Education and Training: The present European Level and Recommendations for its Future Development" [1].
- This policy statement is now considered obsolete and has been replaced by Policy Statement No.12. It remains available on the EFOMP website for reference only.
- Policy Statement 2: "The Roles, Responsibilities and Status of the Clinical Medical Physicist" [3]. This policy statement was also published in 1984 and it concluded that the need for clinical medical physics service in each country depends primarily on the standard and scope of medical care. Generally speaking, it can be said that in the radiological field (X-ray diagnostics, radiotherapy, nuclear medicine and radiation protection), there is an obvious need for a clinical medical physics service. This has proven its development in those countries where the service has been well established. It is also obvious that the introduction of a medical physics service, in general, depends mostly on the appreciation by the medical profession of the ways in which physicists may assist in solving problems of medical diagnosis and treatment.
  The number of physicists per million inhabitants in different European countries shows a wide variation. Figures can be used in comparisons between countries only if they have about the same standard of medical care. Countries striving to reach this standard should in their planning take into account the medical physics service required.
  The number of physicists needed in diagnostic radiology, radiotherapy, nuclear medicine and radiation protection depends on the number of institutions and the number of facilities, such as radiotherapy units. As a rule, countries at an early stage of development of medical physics develop medical radiation physics first, this still being the largest single aspect of a medical physics service.
  EFOMP considers this strategy suitable and that it forms the foundation for the further development of other applications of physics to medical care. The number of clinical medical physicists and supporting staff must be adequate to meet the high standards of service required. In making such provision, health authorities are best guided by the recommendations of the national organizations that are affiliated to EFOMP.
- Policy Statement 3: "Radiation Protection of the Patient in Europe: The Training of the Medical Physicist as a Qualified Expert in Radiophysics" [4].
  This policy statement was published in 1988 in response to Article 5 of the EEC Directive 84/466/Euratom of 3 September 1984 that state "A Qualified Expert in radiophysics shall be available to sophisticated departments of radiotherapy and nuclear medicine".
  In this policy statement, the Qualified Expert in Radiophysics has been defined as "an experienced Medical Physicist working in a hospital, or in a recognised analogous institution, whose knowledge and training in radiation physics are required in services where the quality of the diagnostic image or the precision of treatment is important and the doses delivered to the patients undergoing these medical examinations or treatments must be strictly controlled".
- Policy Statement 4: "Criteria for the Number of Physicists in a Medical Physics Department" [5].
  This policy statement has been revised and has been replaced by Policy Statement number 7. It remains available on the EFOMP website for reference only.
- Policy Statement 5: "Departments of Medical Physics - Advantages, Organisation and Management" [6].
  The recommendations of this policy statement are as follows:
  - The role of Medical Physics Departments is to support the established broad range of applications of physics and engineering in medicine and to be actively involved in the development, implementation and exploitation of new medical technologies and procedures.
  - A main objective of a Medical Physics Department must be to provide a competent and cost-effective medical physics service to all parts of the national health services that need it. This service includes: safety of patients and hospital staff, maintenance of medical equipment and scientific support.

- Medical Physics services must be the responsibility of an integrated Department of Medical Physics providing an agreed core of work activities representative of the diverse character of the specialty.
- These services must be organized or coordinated at the highest practicable level, which can be through a regional or sub-regional structure.
- The Head of the Department must be a physical scientist in medical physics to whom all physical scientists employed on hospital physicists' grades and technical staff must be professionally and officially responsible.
- The Head of the Department should be responsible for the departmental budget.
- University Departments of Medical Physics have the further tasks of teaching, research and training in this field.

• Policy Statement 6: “Recommended guidelines of National Registration Schemes for Medical Physicists” [7].

When EFOMP was inaugurated in May 1980, its principal objective was to harmonise and promote the best practice of medical physics in Europe. In pursuing this objective, one of the long-term aims of EFOMP is to achieve uniformly high standards of training and performance of medical physicists in the countries of all member organisations. Furthermore EFOMP wishes to see some form of recognition when these standards are achieved.

This has three advantages. First, it demonstrates that patients are receiving the same level of medical physics support, no matter where they are being treated. Second, it greatly facilitates the movement of physicists from one country to another. Finally, EFOMP would be seen as the body competent to decide how the by then recently established qualification of European Physicist (Boswell, 1994) would be applied in the context of medical physics.

Within the European Community, the direct application of the Council Directive on "Mutual Recognition of Higher Education Diplomas" 89/48/EEC [8] has not proved a very successful mechanism for ensuring the freedom of movement and the maintenance of appropriate standards for medical physicists. A major reason for this is that the current level of legalised regulation of the profession in Europe is low. However, the European Commission is clearly sympathetic to self-regulation by the professions and in response to an enquiry from FEANI (Fédération Européenne d'Associations Nationales d'Ingénieurs) on the Eur Ing qualification received the reply "The Commission considers that the FEANI scheme is an excellent example of self-regulation by a profession at the European level".

In response to the above, EFOMP provides in this policy statement the necessary guidelines for it to take the lead in establishing a mechanism for the proper recognition of medical physics by means of approved National Registration Schemes.

• Policy Statement 7: “Criteria for the Staffing Levels in a Medical Physics Department” [9].

This policy statement is a revision of Policy Statement number 4 (EFOMP, 1991) and it contains more details as to how many Medical Physicists are required for a Medical Physics Department in relation to the equipment available in the hospital.

Policy statement 7 is now also considered obsolete due to the rapid advancement of the medical technology and it is currently being revised.

• Policy Statement 8: “Continuing Professional Development for the Medical Physicist” [10].

This policy statement was revised in 2000 and it is replaced by Policy Statement number 10.

• Policy Statement 9: “Radiation Protection of the Patient in Europe: The Training of the Medical Physics Expert in Radiation Physics or Radiation Technology” [11].

In order to reach harmonisation throughout Europe when implementing EC-Directive 97/43/Euratom [12] into national legislation, EFOMP recommends the following guidelines with respect to the definition and role of the Medical Physics Expert - MPE:

- At the minimum, the MPE must have been recognised as a qualified medical physicist and preferably also have further experience.

- o The Education and Training Scheme in Medical Physics aiming at the level of a MPE has to follow the EFOMP guidelines [3].
- o A system for a recognised Continuing Professional Development is recommended.
- o According to the duties defined by the new Directive, the MPE has to be involved in radiological practices in all university and specialised hospitals using ionising radiation on patients i.e. radiotherapy, nuclear medicine and diagnostic radiology.
- o Involvement of the MPE in radiological practices, as demanded in Article 6(3) of the Directive, is recommended by EFOMP in the following way:

  EC Directive*:* In radiotherapy, the MPE shall be closely involved.
  EFOMP: Daily relationship between MPE and patient environment is mandatory. To be deeply involved in dosimetry, Quality Assurance and elaboration of techniques used in the radiotherapy department.

  EC Directive*:* In nuclear medicine, the MPE shall be available.
  EFOMP: The MPE shall be able to make a meaningful intervention. Daily relationship between the MPE and the patient environment is most appropriate.

  EC-Directive*:* In diagnostic radiology, the MPE shall be involved as appropriate.
  EFOMP: Depending on the spectrum of techniques used, there must be access to medical physics service, for instance, local or regional networks could be established to provide practitioners and smaller hospitals with up to date medical physics service. When special practices are used, as defined in Article 9 of the Directive, a daily relationship of the MPE and the patient environment shall be standard.

- Policy Statement 10: "Recommended Guidelines on National Schemes for Continuing Professional Development of Medical Physicists" [13].
  This is a revision of Policy Statement number 8 that further develops the recommendations for the establishment of CPD Schemes at the National Level.
- Policy Statement 11: "Guidelines on Professional Contact and Procedures to be implemented in the Event of Alleged Misconduct" [14].
  The role of the medical physicist in health care is diverse. In many areas, the medical physicist will take decisions and give advice that have a direct influence on the management of patients and in all of them medical physicists will interact with individuals from a wide range of professional groups.
  This policy statement gives guidelines on professional conduct that have been drawn up to enable the National Member Organisations of EFOMP to establish a code of practice that will ensure that medical physicists across Europe conduct themselves at all times in a manner that is appropriate to the profession.
- Policy Statement 12: "The present Status of Medical Physics Education and Training in Europe: New Perspectives and EFOMP Recommendations" [2].
  It replaces Policy Statement number 1. The organisation of the Medical Physics Education and Training in many countries has changed since the publication of Policy Statement 1, and this more recent EFOMP policy statement have been issued to introduce new concepts and new recommendations that the current developments in the European Higher Education Area arising from "The Bologna Declaration", and with a view to facilitate the free movement of professionals within Europe, according to the new Directive [15].
  This policy statement provides an updated view of the present level of education and training of Medical Physics in Europe and makes recommendations in view of these new European challenges [16].
- Policy Statement 13: "Recommended Guidelines for the Development of Quality Management Systems for Medical Physics Departments" [17].
  This policy statement outlines the way in which a Quality Management System can be developed for Medical Physics Departments that will be instrumental in eliminating or at least minimizing the contribution of the Medical Physicist to accidents or incidents involving patients. It also describes

what mechanisms need to be in place to ensure the effective and efficient use of new highly complicated and sophisticated technologies and procedures.

2.2. Education and Training Activities

- Support for Meetings, Congresses and Courses*:* The sponsorship of meetings and congresses is instrumental in disseminating and encouraging the adaptation of the policy statements. EFOMP organise, co-organise, support and recognise meetings, congresses and courses with its NMOs. Guidelines that explain the above terms and the requirements for interested NMOs to collaborate in such events can be found on EFOMP's website (www.efomp.org).
  The purpose of these Guidelines is to help NMOs to obtain EFOMP sponsorship for their events by setting out the steps that they need to take and the conditions that must be fulfilled.
  The biggest event is the annual European Conference on Medical Physics. Note that there are detailed guidelines on the requirements for this event.
- Awards and bursaries*:* The encouragement of young scientists to pursue the profession of Medical Physics is of enormous importance to EFOMP. For this reason EFOMP gives awards to young scientist at the EFOMP Conference and provides bursaries to young scientist to attend the European School of Medical Physics – ESMP.
- European School of Medical Physics*:* This is an annual event in collaboration with the European Scientific Institute - ESI and takes place in Archamps, France. It consists of six weeks of intensive training in Medical Physics. One week is spend on covering each of Medical Imaging with Non-ionising Radiation, Medical Imaging with Ionising Radiation, Medical Computing, Physics of Modern Radiotherapy, Brachytherapy, and Radiation Protection.
- European Network of Medical Physics Training Schools*:* This is a new initiative that is currently under development. Moving a step further from just publishing policy statements, the EFOMP Officer's have recently discussed with interested parties on the establishment of a process for coordinating across Europe Continuous Professional Development Courses for Medical Physicists. This would ensure that these are harmonised and offer the same level of education irrespective of the institution delivering them.

The above activities are mainly organised by the Education & Training (ET), the Professional Relations (PR) and the Scientific (SC) Committees of EFOMP.

2.3. Participation in European and International Projects

Presently, EFOMP is participating in the following projects that have within their objectives: the education, training, competence, roles and responsibilities of the Medical Physicist and other Health Professionals working with ionising radiation.

2.3.1. Guidelines on Medical Physics Expert

This is a European Commission project with the objective to provide improved implementation of the Medical Exposures Directive (MED) provisions related to the Medical Physics Expert (MPE) and to facilitate the harmonization of the MPE among Member States aiming at their cross-border mobility.

Up to date information on this project can be obtain from the project's web side at: http://portal.ucm.es/web/medical-physics-expert-project/inicio.

2.3.2. European Medical ALARA Network

This is also a European Commission project and its main objective is to establish a sustainable European Medical ALARA Network (EMAN) where different stakeholders within the medical sector will have the opportunity to discuss and to exchange information on various topics relating to the implementation of the ALARA principle in the medical field.

This network will also support the European Commission (EC) in its activities relating to the optimisation of radiation protection of individuals submitted to medical exposures.

In addition, EMAN will also aim to:

Journal of Physics: Conference Series **317** (2011) 012025 doi:10.1088/1742-6596/317/1/012025

- Disseminate up-to-date information about literature, studies, research and good practices relating to the ALARA principle in the medical sector,
- Identify and communicate to the EC needs for development and update of European Union (EU) guidance,
- In particular cover the areas of education and training as well as continuous quality improvement as requested in the Directive 97/43 EURATOM,
- Formulate proposals to the EC on harmonization issues,
- Propose to the EC solutions of identified issues at the European level,
- Establish co-operation with appropriate international organizations and associations.

Up to date information on this project can be obtain from the project's web side at: http://www.eman-network.eu/.

2.3.3. Medical Radiation Protection Education and Training

The complete title of this project is "Study on the Implementation of the Medical Exposure Directive's Requirements on Radiation Protection Training of Medical Professionals in the European Union".

It is another EC project with the overall aim to provide an improved implementation of the Medical Exposure Directive provisions related to radiation protection education and training of medical professionals in the EU Member States.

The activities of the proposed project will focus on three main tasks:

- The conduction of an EU-wide study on radiation protection training of medical professionals in the EU Member States,
- The organisation of a European Workshop on radiation protection training of medical professionals in the EU Member States,
- The development of a European Guidance document on radiation protection training of medical professionals.

As a result of the European Guidance, the creation of a permanent multidisciplinary working party to draft and maintain European standard sets of competences at various levels for minimum radiation protection training and CPD required for all different groups of medical staff working with ionising radiation shall be envisaged.

Up to date information on this project can be obtain from the project's web side at: http://www.medrapet.eu/.

2.3.4. IAEA Interregional Cooperation Project INT/6/054: "Strengthening Medical Physics in Radiation Medicine

The objective of this IAEA Interregional project is to promote the recognition of medical physics in radiation medicine and to harmonize educational material in order to ensure safe and effective diagnosis and treatment of patients.

More information about this project can be obtained from the projects Technical Officer, Dr. Ahmed Meghzifene (A.Meghzifene@iaea.org or Dosimetry@iaea.org).

4. Conclusions

The above brief discussion describes the activities of EFOMP in the area of Education, Training and Professional Development of the European Medical Physicist.

These activities can only materialise through the collaboration of all the Medical Physicists of EFOMP's NMOs. The NMOs must actively adopt and implement the guidelines of the policy statements as well as participate in the various events organised by EFOMP in collaboration with its NMOs.

The contributions of all interested parties are more than welcome in order to further develop the harmonisation of the education, training and professional status of the Medical Physicist in Europe.

Acknowledgements
EFOMP acknowledges all those that have contributed to the development of its policy statements, all those involved in the organisation of educational and training events under its auspices, as well as those representing EFOMP in European and International projects.

References
[1] EFOMP Policy Statement 1: Medical Physics Education and Training: The present European Level and Recommendations for its Future Development www.efomp.org
[2] EFOMP 2010 Policy Statement 12: The present Status of Medical Physics Education and Training Europe.New Perspectives and EFOMP Recommendations *Phys. Med.* **26** 1-5 www.efomp.org
[3] EFOMP Policy Statement 2: The Roles, Responsibilities and Status of the Clinical Medical Physicist www.efomp.org
[4] EFOMP Policy Statement 3: Radiation Protection of the Patient in Europe: The Training of the Medical Physicist as a Qualified Expert in Radiophysics www.efomp.org
[5] EFOMP Policy Statement 4: Criteria for the Number of Physicists in a Medical Physics Department www.efomp.org
[6] EFOMP 1995 Policy Statement 5: Departments of Medical Physics - Advantages, Organisation and Management *Phys. Med.* **XI**(3) 126-28 www.efomp.org
[7] EFOMP 1995 Policy Statement 6: Recommended guidelines of National Registration Schemes for Medical Physicists *Phys. Med.* **XI**(4) 157-59 www.efomp.org
[8] EEC Council Directive 89/48 1988 General system for the recognition of higher education diplomas awarded on completion of professional education and training of at least three years duration *Off. J. Eur. Comm.* **L32** 16-23.
[9] EFOMP 1997Policy Statement 7: Criteria for the Staffing Levels in a Medical Physics Department *Phys. Med.* **XIII** 187-94 www.efomp.org
[10] EFOMP 1998 Policy Statement 8: Continuing Professional Development for the Medical Physicis *Phys. Med.* **XIV** 81-83 www.efomp.org
[11] EFOMP 1999 Policy Statement 9: Radiation Protection of the Patient in Europe: The Training of the Medical Physics Expert in Radiation Physics or Radiation Technology Phys. Med. **XV** 149-153 www.efomp.org
[12] EC Directive 97/43/Euratom 1997 Health protection of individuals against the dangers of ionising radiation in relation to medical exposures *Off. J. Eur. Comm.* **L180** 22
[13] EFOMP 2001 Policy Statement 10: Recommended Guidelines on National Schemes for Continuing Professional Development of Medical Physicists *Phys. Med.* **XVII** 97-101 www.efomp.org
[14] EFOMP 2003 Policy Statement 11: Guidelines on Professional Contact and Procedures to be Implemented in the Event of Alleged Misconduct *Phys. Med.* **XIX** 227-29 www.efomp.org
[15] Directive 2005/36/EC of the European Parliament and Council 2005 Recognition of professional qualifications *Off. J. Eur. Comm.* **L255** 22-142.
[16] EFOMP The Present Status of Medical Physics Education and Training in Europe: An EFOMP Survey *Phys. Med.* **24** 3-20.
[17] EFOMP 2009 Policy Statement 13: Recommended Guidelines for the Development of Quality Management Systems for Medical Physics Departments *Phys. Med.* **25** 161-165 www.efomp.org

Journal of Physics: Conference Series **317** (2011) 012026 doi:10.1088/1742-6596/317/1/012026

EC Project `GUIDELINES ON MPE´: proposed qualification and curriculum frameworks and the MPE in nuclear medicine

C J Caruana

EFOMP Representative on the EC `Guidelines on MPE´ Project and Biomedical Physics, Faculty of Health Sciences, University of Malta, Malta

carmel.j.caruana@um.edu.mt

`Guidelines on MPE´ Nuclear Medicine workgroup: Alberto Del Guerra - Chairperson (IT), Nicola Belcari (IT), Paola Erba (IT), Stelios Christofides (CY), Carmel J. Caruana (MT), Cesare Gori (IT), Wendy Waddington (UK), Michael Lassmann (DE), Manuel Bardies (FR), Markus Nowak Lonsdale (DK), Bernhard Sattler (DE)

Abstract. The objectives of EC project `Guidelines on Medical Physics Expert´ are to provide for improved implementation of the provisions relating to the Medical Physics Expert within Council Directive 97/43/EURATOM and the proposed recast Basic Safety Standards directive. This includes harmonisation of the mission statement for Medical Physics Services as well as the education and training of the MPE. It also includes detailed knowledge-skills-competence inventories for the Medical Physics Expert in each of Diagnostic and Interventional Radiology, Nuclear Medicine and Radiotherapy. This paper presents the proposed Qualification and Curriculum Frameworks and their application to the Medical Physics Expert in Nuclear Medicine.

1. Introduction

The objective of the European Commission project 'Guidelines on Medical Physics Expert' (TREN/09/NUCL/SI2.549828) is to provide for improved implementation of the provisions relating to the Medical Physics Expert (MPE) of Council Directive 97/43/EURATOM (Medical Exposures Directive, MED) and the recently proposed recast Basic Safety Standards (BSS) Directive. This includes harmonisation of the mission statement and the education and training of the MPE across the Member States of the European Union. In this way, the project will support the European Commission in its actions relating to the optimisation of radiation doses to individuals submitted to medical exposures. This paper presents the proposals of the project consortium on the qualification framework for attainment of MPE status and also for the curricular framework for Education and Training of MPE.

2. Principles guiding the development of the curriculum frameworks

The principles guiding the development of the qualification and curriculum frameworks were the following:

- The proposed qualification framework should be based on the levels defined by the European Qualifications Framework (EQF) for lifelong learning [1] which is the most recent document proposed by the EC for qualification frameworks. For the purpose of this project the appropriate Levels are EQF Level 6 (e.g., Bachelor) and EQF Level 7 (e.g., Masters) [1-2].

- The qualification framework would facilitate the mobility of the MPE in Europe through an agreed set of minimum criteria for achievement of MPE status (whilst keeping in mind the safety of the patient).
- The qualification framework would make it possible for more individuals to achieve MPE status through its flexibility, cost-effectiveness and lifelong learning approach.
- The determination of curriculum content will be guided by a mission statement for Medical Physics Services derived from the MED and recast BSS. The consortium has developed the following mission statement: "Medical Physics Services will contribute to maintaining and improving the quality [3], safety [4-5] and cost-effectiveness [6] of healthcare services through patient-oriented activities requiring expert action, involvement or advice regarding the selection, acceptance, commissioning, quality assurance and optimised clinical use of medical devices used in Diagnostic and Interventional Radiology, Nuclear Medicine and Radiation Oncology and regarding risks from associated physical agents (particularly though not exclusively ionising radiation); all activities will be based on current best evidence or own scientific research when the available evidence is not sufficient". The term 'physical agents' has been used to include both ionising and non-ionising radiation. The use of non-ionising radiation imaging modalities as alternatives to ionising radiation is mandated by Article 3 of 97/43/Euratom and Article 80 of the new BSS. The term 'physical agents' refers to not only ionising radiation but also to static magnetic fields, radiofrequency radiation, ultrasound and any other physical agent associated with these imaging modalities.
  This mission includes the following key activities: General Physics Service, Physical Agents Dosimetry Service, Patient Safety / Risk Management, Occupational and Public Safety / Risk Management (when associated with patient safety), Clinical Medical Device Management, Clinical Involvement, Development of Service Quality and Cost-Effectiveness, Expert Consultancy, Education of Healthcare Professionals and Trainees, Health Technology Assessment (HTA) and Innovation.
- Learning outcomes (LO) for MPE programmes will be expressed in terms of Knowledge, Skills and Competences (KSC) as stipulated and defined in the EQF document as:
  - Knowledge (facts, principles, theories, practices),
  - Skills as the ability to use knowledge and know-how to complete tasks and solve problems (both cognitive skills involving the use of logical, intuitive and creative thinking and practical skills involving manual dexterity and the use of methods, materials, tools and instruments),
  - Competence (in the EQF meaning responsibility and autonomy) [1].
- The curriculum framework must be such as to make it possible for MPE to move easily from one area of medical physics practice (i.e., Diagnostic and Interventional Radiology, Nuclear Medicine and Radiation Oncology) to another, according to national and professional needs and with a minimum amount of additional education and training.

3. Qualification framework for the MPE in Europe

The proposal of the consortium regarding the qualification framework is shown in figure 1.

4. Curriculum framework for MPE programmes in Europe

The curricular framework consists of a structured inventory of LO derived from the key activities of the MPE (listed above) and is shown in figure 2. LO are classified into two categories: Generic and Subject Specific [7].

Generic LO consist of skills which are transferable across professions. They are further classified into the following three sub-categories [7]:

- Instrumental skills - cognitive, methodological, technological, linguistic,
- Interpersonal skills - individual abilities involving social and co-operational aspects,

- Systemic skills - involving combinations of understanding, sensibility and knowledge and usually involving prior acquisition of instrumental and interpersonal skills.

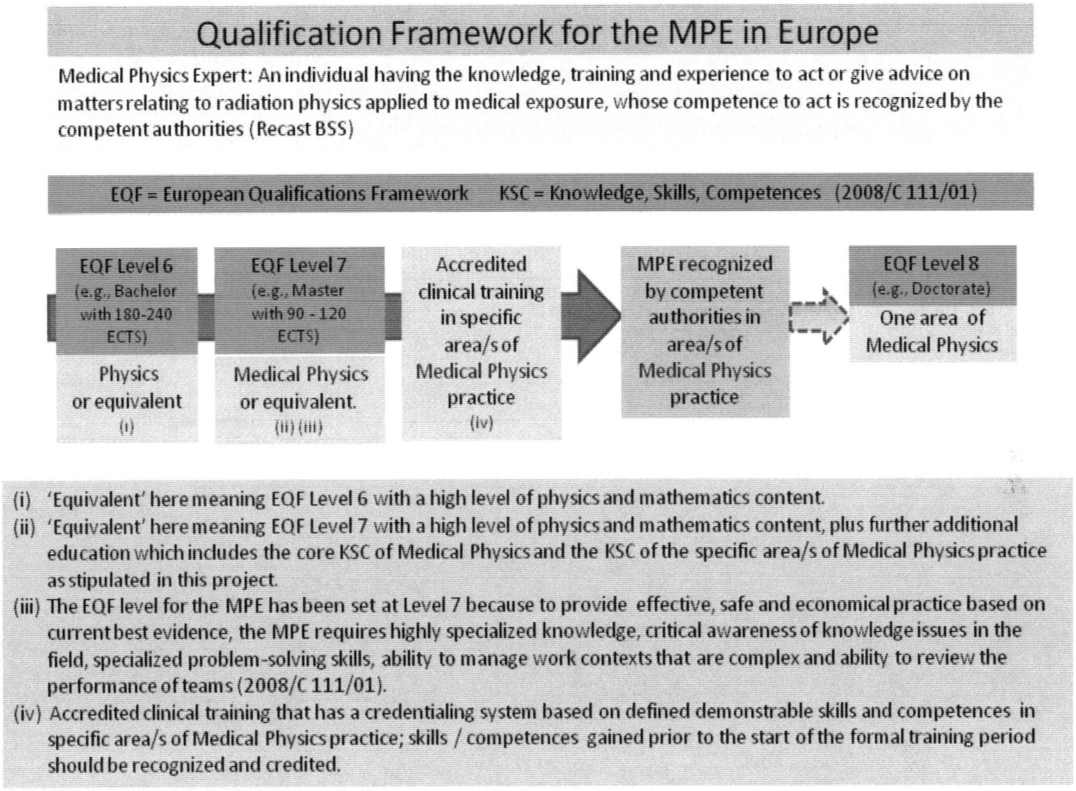

Figure 1. Proposed qualification framework for the MPE in Europe.

Subject Specific LO are those which are specific to a profession. These would be further classified into sub-categories as determined by the particular profession. The following classification is based on proposals by the EFOMP Board and Caruana [8-9]:

- Medical Physics Core KSC. These KSC are expected from all MPE in any area of medical physics practice as follows:
  - KSC appropriate for the MPE as physical scientist (fundamental physics and measurement KSC expected of all physical scientists),
  - KSC for the MPE as healthcare professional (KSC expected of all healthcare professionals),
  - KSC for the MPE as clinical medical devices / physical agents (including ionizing and non-ionizing radiation) expert, representing KSC common to all areas of medical physics practice and reflect the fact that medical physics services are targeted towards the optimised clinical use of medical devices and towards safety from physical agents in the various areas of medicine.
- Medical Physics Areas KSC. These sets of KSC are each highly specific to particular area/s of medical physics practice (i.e., Diagnostic and Interventional Radiology or Nuclear Medicine or Radiation Oncology) and therefore cannot be considered as core.

5. Learning outcomes for Nuclear Medicine

An inventory of LO for the MPE in nuclear medicine was developed by an international group chaired by Prof. A. del Guerra (University of Pisa, Italy). The group reflected on each of the key activities of

Medical Physics Services (listed above) as applied to Nuclear Medicine and derived corresponding LO. Nuclear Medicine LO, which were considered as common to the two other areas of MPE practice, were classified under the core 'Medical Physics Core KSC' grouping. Nuclear Medicine LO, which were highly specific to Nuclear Medicine, were classified under a separate Nuclear Medicine section in the 'Medical Physics Areas KSC' group. The resulting LO inventory is the most comprehensive yet developed. For illustration, table 1 shows the Core KSC for the key activity 'Clinical Involvement', common to the three areas of MPE practice, and table 2, the KSC for the key activity 'Clinical Involvement', which were highly specific to Nuclear Medicine.

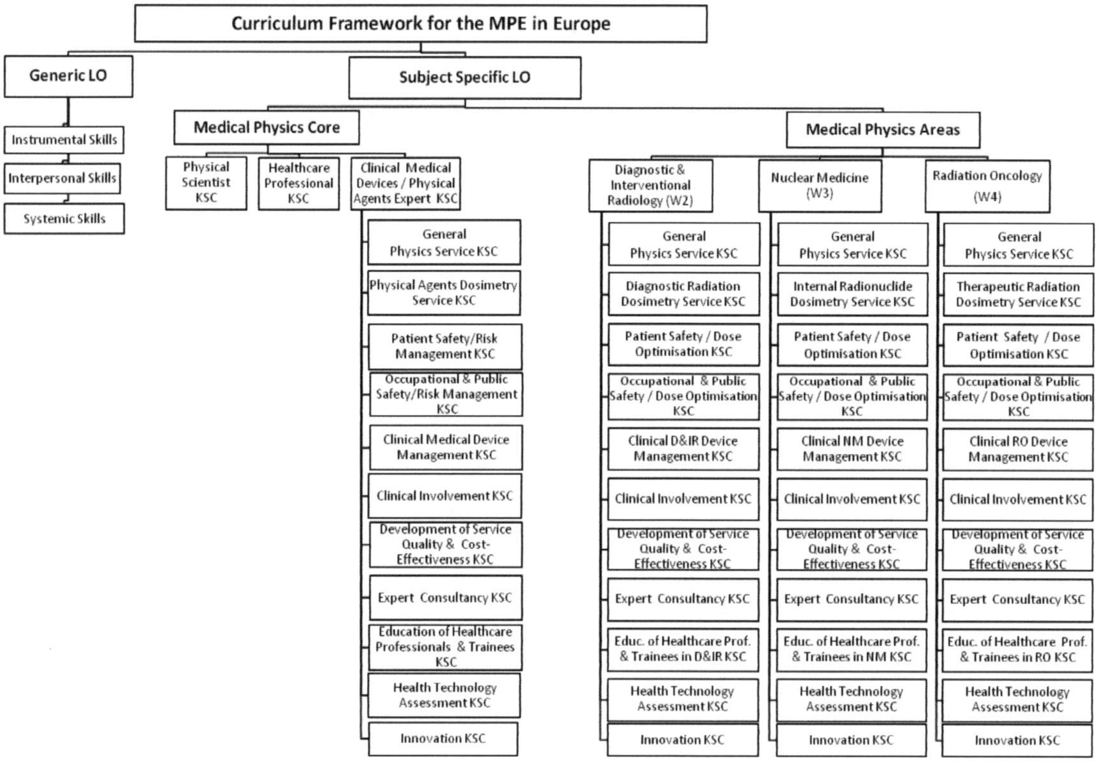

Figure 2. Curriculum framework for MPE programmes in Europe.

6. Outcomes

The Qualification and Curriculum Frameworks have been made available for stakeholder comment on the project website. The website for the project is http://portal.ucm.es/web/medical-physics-expert-project.

Table 1. Core Knowledge-Skills-Competences (KSC) for the key activity 'Clinical Involvement' applicable to all areas of medical physics including nuclear medicine.

Knowledge	Skills	Competences
K1. List and explain statutory and institutional requirements for Medical Physics Services in own area/s of medical physics practice with respect to Clinical Involvement.	S1. Ability to participate in clinical discussions within multidisciplinary teams in own area/s of medical physics practice.	C1. Take responsibility for statutory and institutional requirements for Medical Physics Services in own area/s of medical physics practice with respect to Clinical Involvement.
K2. Describe and explain the principles of anatomy, physiology, biology (including radiobiology), pathology as related to the main clinical applications in own area/s of medical physics practice.	S2. Design patient plans in own area/s of medical physics practice.	C2. Advise on the physics aspects impacting the clinical effectiveness and safety of new medical devices or techniques prior to their introduction into clinical practice.
K3. Describe trauma / development of diseases, diagnosis, treatment and follow-up relevant to own area/s of medical physics practice, including primary healthcare and screening programmes.	S3. Adhere to procedures regarding hygiene.	C3. Advise on and take responsibility for the evaluation and optimization of clinical procedures and protocols and risk elimination / reduction in own area/s of medical physics practice.
K4. Explain the International Classification of Diseases (ICD).	S4. Participate in patient preparation and positioning prior to data acquisition.	C4. Exercise the role of the medical physics expert in clinical involvement within own area/s of medical physics practice.
K5. Explain how medical devices/physical agents are used for the solution of clinical problems in own area/s of medical physics practice.	S5. Analyze critically protocol proposals in terms of feasibility, effectiveness and safety.	C5. Advise physician in imaging interpretation and quantification.
K6. Describe the clinical applications and target clinical outcomes of medical devices/physical agents in own area/s of medical physics practice.	S6. Define the limits of acceptability of clinical procedures.	C6. Take responsibility for deriving semi-quantitative and quantitative data for clinical application.
K7. Describe and explain clinical guidelines in own area/s of medical physics practice.	S7. Assess patient and operator risks for a given experimental procedure.	C7. Advise on different patient diagnosis / treatment schedule options.
K8. Describe the patient's perspective in clinical processes in own area/s of medical physics practice.	S8. Handle and analyze medical images including the extraction of parametric data / images.	C8. Advise on the most appropriate procedure with respect to risk/benefit ratio.
K9. Describe and explain protocol optimization principles in own area/s of medical physics practice.	S9. Set up devices, experiments and protocols for the measurement of physical variables relevant to clinical practice.	C9. Take responsibility for optimization of acquisition protocols in both standard and non-standard situations.
K10. Describe and explain the risk/benefit justification of procedures in own area/s of medical physics practice.	S10. Operate medical devices in own area/s of medical physics practice effectively and safely.	C10. Supervise procedures for paediatric investigations in relation to dose optimisation.
K11. Describe and explain the design principles, the relevant legislation issues and approval procedures for clinical trials.	S11. Recognize basic anatomical / pathological structures of the human body in projection /	C11. Advise other healthcare
K12. Explain the principles and implementation of Good Clinical Practice (GCP), Good Manufacturing Practice (GMP) and Good Laboratory Practice (GLP) in own area/s of medical physics practice.		
K13. Describe general indications and contra-indications for the use of devices in own area/s of medical physics practice.		
K14. Understand the nature of anatomical/ pathological medical images as the visualization of the 3D distribution of physical variables.		

K15. List the main sources of evidence from within the general physics, medical physics and general healthcare (e.g., the Cochrane Collaboration) literature essential for the carrying out of a systematic survey in own area/s of medical physics practice. K16. Explain basic concepts in health informatics such as unique patient identifier, medical record and disease coding (e.g., ICD10). K17. Explain safety and risk related issues associated with the use of ICT in own area/s of medical physics practice.	tomographic and 3D medical images relevant to own area/s of medical physics practice. S12. Recognize basic physiological processes in nuclear / molecular images.	professionals on optimization and safety of individual patient examination/treatment and examination/treatment protocols. C12. Live up to demands imposed by duty of confidentiality, professional secrecy, ethical standards. C13. Represent medical physics in clinical conferences.

Table 2. KSC for the key activity 'Clinical Involvement' specific to Nuclear Medicine.

Knowledge	Skills	Competences
K1. List statutory and institutional requirements for Medical Physics Services in Nuclear Medicine with respect to Clinical Involvement. K2. Describe the principles of anatomy, physiology, biology, radiobiology and pathology as related to the main clinical application of Nuclear Medicine diagnostic and therapeutic techniques. K3. Describe the general role of Nuclear Medicine procedures in diagnosis, therapy and treatment response evaluation. K4. Explain how Nuclear Medicine devices are used for the solution of a clinical problem. K5. Describe the principle of radiopharmaceutical preparation and associated quality control. K6. Describe the principles of radiopharmaceutical biodistribution in normal organ and target tissues. K7. Describe the fundamentals of molecular radiotherapy. K8. Explain the fundamentals of the use of Nuclear Medicine procedures in EBRT planning. K9. Describe general indications and contra-indications for Nuclear Medicine procedures. K10. Describe diagnostic and clinical procedure guidelines. K11. Describe protocol optimization principles. K12. Describe the risk/benefit justification of Nuclear Medicine diagnostic and therapeutic procedures as related to radiation exposure risk. K13. Explain the interactions/synergism between chemotherapy, EBRT and molecular radiotherapy.	S1. Develop optimized imaging and therapeutic protocols. S2. Design a treatment plan. S3. Show how Nuclear Medicine data are used in radiotherapy planning. S4. Analyze how molecular radiotherapy could impact on other treatment modalities. S5. Analyze critically protocol proposal (i.e. feasibility, safety…). S6. Analyze the limits of acceptability of clinical Nuclear Medicine procedures. S7. Calculate patients and operators doses	C1. Take responsibility for statutory and institutional requirements for Medical Physics Services in Nuclear Medicine with respect to Clinical Involvement. C2. Advise Nuclear Medicine physicians in imaging interpretation and quantification. C3. Take responsibility for deriving semi-quantitative and quantitative data for clinical application. C4. Advise on different treatment schedule options. C5. Advise on the most appropriate procedure with respect to risk/benefit ratio. C6. Advise on and take responsibility for optimization of clinical acquisition protocols in both standard and non-standard situations. C7. Supervise procedures for paediatric investigations. C8. Evaluate clinical trial protocols. C9. Share responsibility for conducting clinical trials. C10. Advise on relevant aspects of ethical review of a clinical trial C11. Assume responsibility for data handling / recording.

K14. Illustrate methodologies for the measurement of the lesions response to therapy. K15. List laboratory and imaging procedures to evaluate organ toxicity. K16. Illustrate dose limiting toxicity classification and quantification. K17. Describe and explain the design principles, the relevant legislation issues and approval procedures for clinical trials in Nuclear Medicine.	and consequent risks for a given clinical or experimental procedure.	C12. Advise on and take responsibility for the optimization of clinical protocols for Nuclear Medicine equipment including software. C13. Support Nuclear Medicine staff with physical-technical guidelines.

Acknowledgment
The members of the consortium would like to thank the European Commission for providing the funds to make this project possible.

References

[1] European Parliament and Council Recommendation 2008/C 111/01 on the establishment of the European Qualifications Framework for Lifelong Learning. *Official Journal of the European Union* 6.5.2008. http://eurlex.europa.eu/LexUriServ/LexUriServ.do?uri=oj:c:2008:111:0001:0007:en:pdf
[2] A Framework for Qualifications of the European Higher Education Area http://www.bologna-bergen2005.no/Docs/00-Main_doc/050218_QF_EHEA.pdf
[3] Legido-Quigley H, McKee M, Nolte E and Glinos I A 2008 *Assuring the quality of health care in the European Union* http://www.euro.who.int/__data/assets/pdf_file/0007/98233/E91397.pdf
[4] Luxembourg Declaration on Patient Safety - *Making it Happen* http://www.eu2005.lu/en/actualites/documents_travail/2005/04/06Patientsafety/Luxembourg_Declaration_on_Patient_Safety_05042005-1.pdf
[5] Standing Committee of European Doctors and other stakeholders 2005 *Stakeholders' position paper on patient safety* (http://www.ehtel.org/forum/tasks-sources/tf-patient-safety-emedication/stakeholders_position_on_patient_safety.pdf)
[6] EC 2007 Together for Health: A Strategic Approach for the EU 2008-2013 http://eur-lex.europa.eu/LexUriServ/site/en/com/2007/com2007_0630en01.pdf
[7] Tuning - Universities' contribution to the Bologna Process 2nd ed. http://www.tuning.uni deusto.org/tuningeu/
[8] Christofides S (EFOMP Officers) and Caruana C J 2009 An Initial EFOMP Position on the Tuning Process for Masters Programs in Medical Physics in Europe *IFMBE Proc. 25/XIII 2009* (Munich) Editors: Dössel O and Schlegel W C (Berlin: Springer) pp. 70–73
[9] Caruana C J 2010 The Tuning process and the Masters in Medical Physics in Europe - a personal vision *Medical Physics and Engineering - Education and Training*. Editors: Tabakov S, Sprawls P, Krisanachinda A, Lewis C. (Trieste: ICTP Press) (in press).

AUTHOR INDEX